ADAPTING TO CLIMATE CHANGE IN EASTERN EUROPE AND CENTRAL ASIA

Contents

Boxes

Figures

Maps

Tables

About the Editors and Authors

Tereen Alireza, formerly a World Bank consultant, is studying in the Environment, Society, and Development program at Cambridge University. She has an MSc in Environmental Science from the University of Sydney.

Rachel I. Block has worked as a consultant in ECA and for the *World Development Report 2010* on climate change. She holds an MSc in Economics from Universitat Pompeu Fabra in Barcelona.

JoAnn Carmin is Associate Professor in the Department of Urban Studies and Planning at the Massachusetts Institute of Technology. She holds a PhD in Environmental Policy and Planning from the University of North Carolina at Chapel Hill.

Timothy Carrington, a writer and consultant with 15 years of experience at the *Wall Street Journal,* was previously a senior communications officer and manager of a journalist training program at the World Bank. He has a BA from the University of Virginia.

Nicola Cenacchi, a biologist with an MSc in Environmental Technology from Imperial College London, is a consultant in the Sustainable Development department in ECA. He also contributed to the *World Development Report 2010* on climate change.

Jane Ebinger is a Senior Energy Specialist and Thematic Coordinator of Energy and Climate Change at the World Bank. She holds an MA in Mathematics and an MSc in Mathematical Modeling and Numerical Analysis from Oxford University.

Barbara Evans is a Senior Lecturer at the School of Engineering, University of Leeds, following nearly 15 years working on water and sanitation at the World Bank. She holds a BEng in Civil Engineering from the University of Leeds and an MSc in Development Studies from the London School of Economics.

Marianne Fay, Chief Economist of the World Bank's Sustainable Development Network, was formerly a Lead Economist in ECA and the Director of the *World Development Report 2010*. She holds a PhD in Economics from Columbia University.

Katharina Ferl, formerly a consultant the Human Development Sector in ECA, is at the Research Department of the International Monetary Fund. She holds an MA from the Johns Hopkins School of Advanced International Studies.

Franz Gerner, a Senior Energy Economist in ECA, focuses on the World Bank's gas dialogue in Southeastern Europe and Central Asia and leads two national energy investment portfolios. He holds a PhD in Economics from the University of Innsbruck.

Bjorn Hamso is a Senior Energy Economist, formerly in ECA and now in South Asia, and has worked previously in the Norwegian oil and gas industry. He has an MSc from the Norwegian School of Economics and Business Administration.

Vladimir Kattsov is the Director of the Voeikov Main Geophysical Observatory (MGO) in St. Petersburg. He attained a Doctor of Physical and Mathematical Sciences from Voeikov MGO and a PhD in oceanology from Leningrad Hydrometeorological Institute.

Alexey Kokorin is the Climate Change Program Coordinator at WWF Russia. He has a Doctor of Sciences degree from the Russian Academy of Sciences Institute of Global Climate and Ecology and a PhD from Moscow State University.

Jolanta Kryspin-Watson is an Operations Officer at the World Bank with a decade of experience specifically in disaster recovery operations and risk mitigation. She has an MBA from the University of Warsaw and an MPA from the State University of New York, Albany.

Antonio Lim was an Operations Officer who worked in ECA for 15 years on health, social safety nets, and energy. He did graduate studies in economics at the University of the Philippines, the University of Sussex, and American University.

Ziad Nakat, now in the Middle East and North Africa region, is a Transport Specialist with the Energy and Transport Unit. He holds an MA in Economics, and an MS and a PhD in Civil and Environmental Engineering, from the University of California, Berkeley.

Sonja Nieuwejaar is a consultant and expert on disaster management. She worked as a Deputy Director in the Office of International Affairs of the United States Federal Emergency Management Agency (FEMA). She holds an MA in International Relations from Yale University.

Ana Plecas, formerly a consultant with the energy team in ECA, is with the South Asia transport team. She holds an MA in International Affairs from the Johns Hopkins School of Advanced International Studies.

John Pollner is a Lead Financial Officer in the ECA region, specializing in insurance, disaster risk, and financial sector management. He holds an MBA from the University of California, Berkeley.

Tamer Rabie is a Senior Health Specialist in ECA. He holds an MBBCh from Cairo University and an MSc in Public Health from the London School of Hygiene and Tropical Medicine, where he is pursuing his PhD in Epidemiology.

Gerardo Sanchez worked on environment, health, and sustainable urbanization at the World Bank and is now a Technical Officer in Urbanization and Emergency Preparedness at the World Health Organization. He holds an MPH from Boston University School of Public Health.

Jitendra P. Srivastava, now a consultant, is an agronomist with 30 years of experience at the World Bank and over 20 years in ECA. He has specialized in drought tolerance and agricultural research and training. He holds a PhD from the University of Saskatchewan.

William R. Sutton is a Senior Agricultural and Resource Economist in ECA working on rural development and the integration of environmental concerns. He holds a PhD in Agricultural and Resource Economics from the University of California, Irvine.

Safinaz el Tahir is a Rural Development Specialist in ECA. She has worked for the World Food Programme and Save the Children USA. She holds an MSc in Development Studies from the London School of Economics.

Michael Webster is a Senior Water and Sanitation Specialist in ECA. He holds an MSc in Water and Waste Engineering from Loughborough University and a Masters in Public Policy from Princeton University.

Michael I. Westphal is a consultant specializing in conservation and ecological modeling. A former Science Policy Fellow for the American Association for the Advancement of Science, he holds a PhD from the University of California, Berkeley, in Environmental Science, Policy, and Management.

Yan F. Zhang is a Senior Urban Economist in ECA, focusing on housing, land reforms, and fiscal decentralization. She has a Masters in Urban Planning from the China Academy of Urban Planning and Design, and a PhD from the Department of Urban Studies and Planning at the Massachusetts Institute of Technology.

Acknowledgments

This Report is dedicated to the memory of contributor Antonio Lim, whose kindness and diligence have impacted the many clients with whom he worked in the ECA region, as well as all of us who came to know him during the writing of this Report and throughout his 15 years at the World Bank.

This publication was produced by the office of the Chief Economist of the Europe and Central Asia (ECA) Region of the World Bank. It draws from the work of a team that produced the following background papers (all of which are available at http://www.worldbank.org/eca/climatechange):

On climate science and vulnerability
- *Summary of the Climate Science in the Europe and Central Asia Region: Historical Trends and Future Projections* by Michael I. Westphal
- *A Simple Index Of Vulnerability to Climate Change* by Marianne Fay and Hrishi Patel

On the natural environment and agriculture
- *Adaptation to Climate Change in Coastal Areas of the ECA Region* by Nicola Cenacchi
- *Biodiversity Adaptation to Climate Change in the ECA Region* by Nicola Cenacchi
- *Adaptation to Climate Change in Europe and Central Asia Agriculture* by William R. Sutton, Rachel I. Block, and Jitendra P. Srivastava

On infrastructure

- *Adapting to Climate Change in Europe and Central Asia; Background Paper on Water Supply and Sanitation,* by Barbara Evans and Michael Webster
- *Achieving Urban Climate Adaptation in Europe and Central Asia,* by JoAnn Carmin and Yan F. Zhang
- *Europe and Central Asia Region: How Resilient is the Energy Sector to Climate Change?* by Jane Ebinger, Bjorn Hamso, Franz Gerner, Antonio Lim, and Ana Plecas
- *Climate Change Adaptation in the Transport Sector* by Ziad Nakat

On health

- *The Health Dimension of Climate Change* by Tamer Rabie, Safinaz el Tahir, Tereen Alireza, Gerardo Sanchez, Katharina Ferl, and Nicola Cenacchi

On disaster management

- *Climate Change Adaptation in Europe and Central Asia: Disaster Risk Management* by John Pollner, Jolanta Kryspin-Watson, and Sonja Nieuwejaar

On Russia and Central Asia

- *Climate Change Projections and Impacts in the Russian Federation and Central Asia* by Vladimir Kattsov, Veronika Govorkova, Valentin Meleshko, Tatyana Pavlova, Igor Shkol
- *Expected Impact of the Changing Climate on Russia and Central Asia Countries* by Alexey Kokorin (WWF Russia)
- *Ongoing or Planned Adaptation Efforts and Strategies in Russia and Central Asia Countries* by Alexey Kokorin (WWF Russia)

In addition, we are grateful for the work of Tim Carrington, who drafted significant portions of the final report based on these papers, and for the inputs of the following individuals: Shelley McMillan and Irene Bomani (water resource management) and Andrea Liverani (barriers to adaptation). Hrishi Patel produced the vulnerability indicators, building on earlier work by Elena Strukova and Muthukumara S. Mani.

We would also like to thank Vladimir Tsirkunov and Lucy Hancock, who generously shared the experience and information obtained while working on the hydromet study that is featured in chapter 7, and Elena Kantarovich, who was critical in managing the work of such a large team.

The report benefited from comments from peer reviewers Ian Noble (ENV) and Chris West (UKCIP), and from Salman Anees (ECAVP), Luca Barbone (ECSPE), Ron Hoffer (ECSSD), Erika Jorgensen (ECSPE), Orsalia Kalantzopoulos (ECCU5), and John Nash (LCSSD). The work was conducted under the general guidance of Pradeep Mitra, followed by Indermit Gill, in the role of Chief Economist in ECA.

ECA Countries and Subregions

For this book, the authors chose the regional groupings of the ECA countries based on current climate, projected climate, and general economic and agricultural characteristics. Because of its vast and varied territory, the Russian Federation has been divided into six subregions based on climate and the geographic distribution of agricultural activity.

Regional groupings	Economies
Southeastern Europe	Albania, Bosnia and Herzegovina, Bulgaria, Croatia, FYR Macedonia, Montenegro, Kosovo, Serbia, Slovenia, Turkey
Central and Eastern Europe	Czech Republic, Hungary, Moldova, Romania, Slovak Republic, Ukraine
Baltics	Belarus, Poland, Estonia, Latvia, Lithuania
South Caucasus	Armenia, Azerbaijan, Georgia
Central Asia	Tajikistan, Uzbekistan, Turkmenistan, Kyrgyz Republic
Kazakhstan	Kazakhstan

Russia subregions	Oblasts, republics, and districts
Baltic and Western Arctic	Arkhangelsk, Kaliningrad, Karelia, Komi, Kostroma, Leningrad, Murmansk, Nenetsk, Novgorod, Pskov, St Petersburg, Tver, Vologda, Yamalo-Nenetsk, Yaroslavl, Arctic Islands
Central and Volga	Bashkortostan, Belgorod, Bryansk, Chuvashia, Ivanovo, Kaluga, Kirov, Komi-Permyak, Kursk, Lipetsk, Mari El, Mordovia, Moscow, Nizhniy Novgorod, Orel, Orenburg, Penza, Perm, Ryazan, Samara, Saratov, Smolensk, Tambov, Tatarstan, Tula, Udmurtia, Ulyanovsk, Vladimir, Volgograd, Voronezh
North Caucasus	Adygea, Astrakhan, Chechnya, Dagestan, Ingush, Kabardino-Balkaria, Kalmykia-Khalmg Tan, Karachay-Cherkessia, Krasnodar, North Ossetia-Alani, Rostov, Stavropol
Urals and Western Siberia	Altay, Chelyabinsk, Kemerovo, Khakassia, Khant-Mansiysk, Kurgan, Novosibirsk, Omsk, Sverdlovsk, Tomsk, Tyumen, Tyva
South Siberia	Aga Buryatia, Amur, Buryatia, Chita, Irkutsk, Ust-Orda Buryat
East Siberia and the Far East	Chukotka, Evenk, Jewish, Kamchatka, Khabarovsk, Koryak, Krasnoyarsk, Magadan, Primorskiy, Sakha-Sakhalin, Arctic Islands

Executive Summary

The climate is changing, and ECA is already experiencing the consequences: increasing variability, warmer temperatures, changing hydrology, and more extremes—droughts, floods, heat waves, windstorms, and forest fires.

With a legacy of environmental mismanagement and underinvestment in infrastructure and housing, the region is already vulnerable to the current climate conditions because of its "adaptation deficit," which can only increase with projected climate changes. In the near term, the region's vulnerability is dominated by non-climatic factors, including socioeconomic and environmental issues that are the legacy of the former Soviet system. These will exacerbate climate risks and hamper the ability of sectors that could gain from climate change, such as agriculture, to reap full benefits.

Certainty about global warming and the dismal consequences of unmitigated emissions coexist with uncertainty about local impacts and the timing of particular weather events. Policy makers at national and local levels, individuals, and business owners may face substantial uncertainty as to what to adapt to. The focus therefore must not be on precise impact assessment, but on reducing vulnerability, starting with vulnerability to the current climate. Postponing action until more is known would be a mistake. It would also preclude taking advantage of the many opportunities to increase resilience while reaping copious co-benefits.

This book has four key messages:

- **Contrary to popular perception, ECA faces a substantial threat from climate change, with a number of the most serious risks already in evidence.** Average temperatures across ECA have already increased by 0.5°C in the south to 1.6°C in the north (Siberia), and overall increases of 1.6 to 2.6°C are expected by the middle of the century regardless of what mitigation efforts are undertaken. This is affecting hydrology, with a rapid melting of the region's glaciers and a decrease in winter snows. Many countries are already suffering from winter floods and summer droughts— with both Southeastern Europe and Central Asia at risk for severe water shortages. Summer heat waves are expected to claim more lives than will be saved by warmer winters.

- **Vulnerability over the next 10 to 20 years will be dominated by socioeconomic factors and legacy issues**—notably the dire environmental situation and the poor state of infrastructure— rather than by the changing climate itself.

- **Even countries and sectors that stand to benefit from climate change are poorly positioned to do so.** Many have claimed that warmer climate and abundant precipitation in the northeastern part of ECA (Kazakhstan, Russia, and Ukraine) will open up a new agricultural frontier. However, the region's currently low agricultural performance, with efficiency and productivity levels far below those of western Europe, does not augur well for its capacity to seize new opportunities.

- **The next decade offers a window for ECA countries to make their development more resilient to climate change while reaping numerous co-benefits.** While some impacts of climate change are already being felt, they will likely remain manageable over the next decade, thereby offering the ECA region a short period of time to increase its resilience by focusing on actions that have numerous co-benefits.

More details on particular sectors or countries can be found in the numerous background papers that underpin the report at http://www.worldbank.org/eca/climatechange.

Abbreviations

CDD	consecutive dry days
CO_2	carbon dioxide
ECA	Eastern Europe and Central Asia
EM-DAT	Emergency Events Database
EU	European Union
GCM	General Circulation Model
GDP	gross domestic product
GHG	greenhouse gas
IPCC	Intergovernmental Panel on Climate Change
IWRM	Integrated Water Resource Management
km^2	square kilometer
km^3/y	cubic kilometer per year
ktoe	kiloton of oil equivalent
m	meter
OECD	Organisation for Economic Co-operation and Development
PRUDENCE	Prediction of Regional scenarios and Uncertainties for Defining EuropeaN Climate change risks and Effects
RCM	Regional Climate Model
SWIFT	Structured What If Technique
UKCIP	United Kingdom Climate Impacts Programme
WMO	World Meteorological Organization
WWF	World Wide Fund For Nature

Overview

The climate is changing, and the Eastern Europe and Central Asia (ECA) region is vulnerable to the consequences.[1] Many of the region's countries are facing warmer temperatures, a changing hydrology, and more extremes—droughts, floods, heat waves, windstorms, and forest fires. Already the frequency and cost of natural disasters in the region have risen dramatically. And the concentration of greenhouse gases already in the atmosphere guarantees that equally large or even larger changes are yet to come—even if the world completely stopped emitting CO_2.

Yet, for the near future, ECA's vulnerability is driven more by its existing sensitivity than by the severity of the likely climate impacts. In fact, ECA already suffers from a serious adaptation deficit even to its current climate. This derives from a combination of socioeconomic factors and the Soviet legacy of environmental mismanagement. Such mismanagement is perhaps the most dangerous holdover, massively increasing vulnerability to even modest global warming. Thus, the expected decrease in the level of the Caspian Sea means that the population will come into contact with a range of dangerous substances such as pesticides and arsenic that are presently locked in coastal sediments. Rising temperatures and reduced precipitation in Central Asia will exacerbate the environmental catastrophe of the disappearing Aral Sea.

The region also bears the burden of poorly constructed, badly maintained, and aging infrastructure and housing—a legacy of both the Soviet era and the transition years. These are ill-suited to cope with storms, heat waves, or floods, much less to protect populations from the impacts of such extreme events. While Turkey does not carry the same legacy issues, it suffers from demographic pressures on fragile natural resources and inadequate and vulnerable infrastructure.

This book has four key messages:

- *Contrary to popular perception, Eastern Europe and Central Asia face significant threats from climate change, with a number of the most serious risks already in evidence.* Average temperatures across ECA have already increased by 0.5°C in the south to 1.6°C in the north, and overall increases of 1.6 to 2.6°C are expected by the middle of the century. This is affecting hydrology, with a rapid melting of the region's glaciers and a decrease in winter snows. Many countries are already suffering from winter floods and summer droughts—with both Southeastern Europe and Central Asia at risk for severe water shortages. Summer heat waves are expected to claim more lives than will be saved by warmer winters.

- *Vulnerability over the next 10 to 20 years is likely to be dominated by socioeconomic factors and legacy issues*—notably the dire environmental situation and the poor state of infrastructure—rather than by the changing climate itself. A flood in Baia Mare, Romania, in 2000 brought cyanide-laced waste from a gold mining operation into the Tiza and Danube Rivers, poisoning the water of 2 million people. In subregions threatened with water shortages, poor water management dwarfs the climate change impacts anticipated for the next 20 years.

- *Even countries and sectors that stand to benefit from climate change are poorly positioned to do so.* Warmer climate and abundant precipitation in the north-central part of ECA (Kazakhstan, Russian Federation, and Ukraine) could open up a new agricultural frontier. However, any potential benefit pales in comparison to the costs of the region's relative inefficiency and low productivity. While world grain yields have been growing on average by about 1.5 percent per year, they have been falling or stagnant in these three countries, where productivity is far below that of Western Europe or the United States.

- *The next decade offers a window of opportunity for ECA countries to make their development more resilient to climate change while reaping numerous co-benefits.* While some impacts of climate change are already being felt, these will likely remain manageable over the next decade. This offers the ECA region a short period to increase its resilience by focusing on "no-regret" beneficial actions. Regardless of climate change, ECA will gain a lot by improving its water resource management, fixing its disastrous environmental legacy, upgrading neglected infrastructure and housing, and strengthening disaster management. But the region should also develop strategies to reduce vulnerability to future changes—focusing on infrastructure but also capacity-building and stronger institutions to support adaptation. And forward-looking decisions today help avoid locking countries or settlements into unsustainable patterns of development. Experiences from other countries, regions, or cities now developing and implementing adaptation plans offer valuable lessons and methodologies

This book presents an overview of what adaptation to climate change might mean for Eastern Europe and Central Asia. It starts with a discussion of emerging best-practice adaptation planning around the world and a review of the latest climate projections. It then discusses possible actions to improve resilience organized around impacts on health, natural resources (water, biodiversity, and the coastal environment), the "unbuilt" environment (agriculture and forestry), and the built environment (infrastructure and housing). The last chapter concludes with a discussion of two areas in great need of strengthening given the changing climate: disaster preparedness and hydrometeorological services.

Adaptation to climate change is a nascent field, much less studied and understood than mitigation, which describes actions to reduce emissions of greenhouse gases. Although adaptation was not part of the Kyoto negotiations in 1992, it now stands as one of the four pillars (with mitigation, finance, and technology) of the negotiations underway within the United Nations Framework Convention on Climate Change, which will culminate at the Climate Change Conference in Copenhagen in December 2009. Hence the focus of this book is on adaptation, as opposed to mitigation, which is addressed in a number of other World Bank projects and reports.

Climate Change—A Major Threat to Eastern Europe and Central Asia

Both temperatures and precipitation are projected to change significantly over the coming decades in the ECA countries. Temperatures will continue increasing everywhere in the region, with the greater changes occurring in the more northern latitudes. The north is projected to see its greatest temperature changes in winter, while southern parts of the region are expected to see more warming in summer than in winter. Overall in the region, the number of frost days is projected to decline by 14 to 30 days over the next 20 to 40 years (see map 2.2), with the number of hot days increasing by 22 to 37 days over the same period. This warming trend is significant: by mid-century, countries such as Hungary or Poland are expected to experience the same number of hot days (>30°C) as today's Spain or Sicily.

Water availability is projected to decrease everywhere but Russia (see map 2.4b), as increased precipitation in many regions is offset by greater evaporation due to higher temperatures. The most dramatic decreases are likely to occur in Southeastern Europe (–25 percent), where, along with parts of Central Asia, the absolute amount of precipitation will decline. Even in Russia, most of the precipitation increase is expected to occur in winter; therefore, it is still possible that higher summer temperatures could offset precipitation and lead to drought conditions.

Yet even as much of the region is faced with possible droughts, floods are expected to become more common and severe. This is because precipitation intensity will increase across the region—notably through more frequent storms. Although models cannot predict floods per se—as they are events brought on by many factors other than precipitation, such as land use—ECA is in fact already experiencing more severe and frequent floods. Without substantial adaptation measures, the new weather pattern is likely to result in more floods.

Warmer temperatures also mean that glaciers are receding and that less winter precipitation falls and is stored in the form of snow. This complicates hydrology and makes it more likely for ECA to experience more winter flooding. And while in the short term basins that rely on glacial melt for summer water may see increased water flow from melting glaciers, the long-term implications for summer water availability are troubling—particularly in irrigation-dependent Central Asia.

In the Arctic, temperatures have been warming at about twice the global average with significant impacts on arctic ice, the tundra, and permafrost. Ice cover in September (when the ice is at its minimum) is projected to decline 40 percent by mid century. Some models project that by the end of the century the Arctic will be completely ice-free in the summer. Russia's permafrost line is receding, and seasonal thaw depths are projected to increase by 30 to 50 percent by 2050. The melting of ice and permafrost is affecting biodiversity, as well as leading to coastal erosion and the collapse of exposed buildings and infrastructure.

Changes in sea level, another impact of climate change, will affect ECA's four major basins (the Baltic Sea, the East Adriatic and Turkey's

Mediterranean coast, the Black Sea, and the Caspian) and the Russian Arctic Ocean. On the Baltic, Poland's heavily populated low-lying coast is especially vulnerable to sea-level rise. Along the Adriatic and the Mediterranean, storm surge and saltwater intrusion into aquifers threaten parts of the Albanian, Croatian, and Turkish coasts. Sea-level rise has been highest in the Black Sea, where it is threatening the numerous ports and towns along the Georgian, Russian, and Ukrainian coasts. In the Caspian Sea, water levels are projected to drop by approximately six meters by the end of the twenty-first century, due to increased surface evaporation. This will imperil fish stocks and affect coastal infrastructure.

An index designed to capture the strength of future climate change relative to today's natural variability (Baettig, Wild, and Imboden 2007) suggests that the ECA countries most exposed to increased climate extremes are, relative to the rest of the world, in the middle tier of exposure, and include Russia, Albania, and Turkey, followed by Armenia, the Former Yugoslav Republic of Macedonia, and Tajikistan (figure 1). Relative to the rest of the world, these countries are in the

FIGURE 1

ECA Countries Likely to Experience the Greatest Increases in Climate Extremes by the End of the 21st Century: Russia, Albania, and Turkey

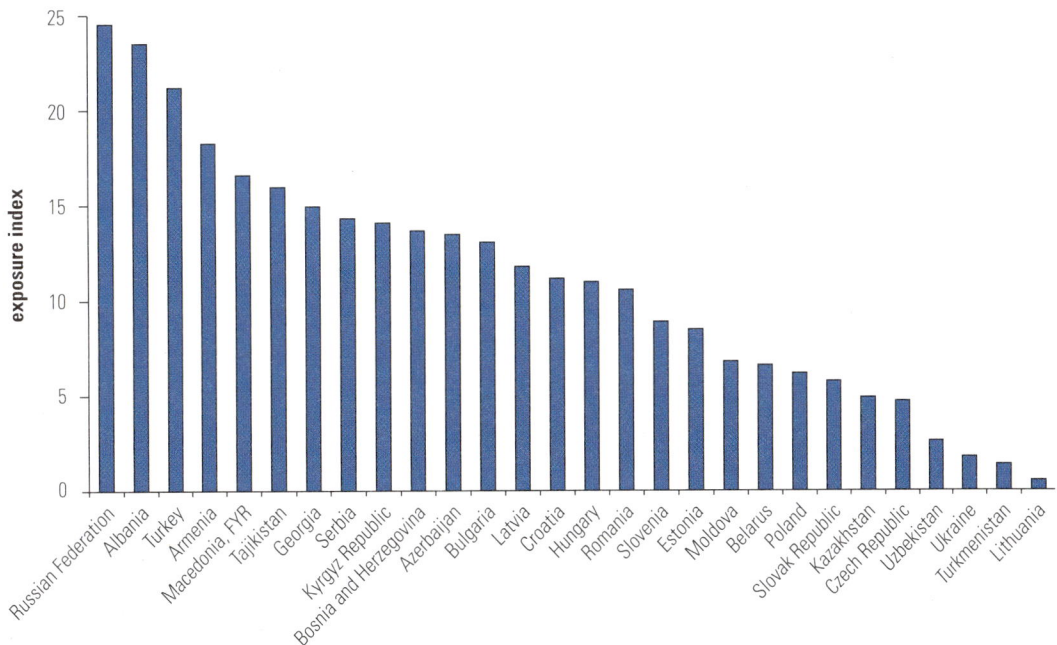

Source: Baettig, Wild, and Imboden 2007.

Note: The index combines the number of additional hot, dry, and wet years; hot, dry, and wet summers; and hot, dry, and wet winters projected over the 2070–2100 period relative to the 1961–90 period. As such, countries already experiencing substantial variability and extremes are less likely to rank highly on this index; for example, India—with large projected variability but already high variability today—and the Czech Republic have about the same score.

middle tier of exposure. However, this ordering is not necessarily reflected in a concern for climate change: only 40 percent of Russians think climate change is a serious issue; in contrast, 70 percent of Turks do (Pew Global Attitudes Project 2007; see figure 1.7).

Increased temperatures and changing hydrology are already affecting ECA's forestry and agriculture. Extreme events combined with earlier snowmelt and hot, dry summers have caused substantial tree loss and degradation. In Russia, 20 million hectares of forest were lost to fire in 2003 alone. The warming climate is also allowing the northward migration of pests and harmful plant species. For agriculture, net losses are likely for Southeastern Europe and Turkey, the North and South Caucasus, and Central Asia. The projected impacts are mixed or uncertain in Central and Eastern Europe, Kazakhstan, and the Central and Volga region of Russia.

Warmer weather and other factors associated with climate change are also affecting health. Malaria, which had been eradicated from Europe, is making a comeback, as are other once-rare infectious diseases; meanwhile, allergies related to pollen are projected to increase, particularly in Central Europe. Hundreds of deaths were attributed to the 2001 heat waves in Moscow and across Croatia, the Czech Republic, and Slovenia. Such heat waves will occur much more frequently in the future.

Vulnerability Will Be Dominated by Socioeconomic Factors and Legacy Issues

Resilience to a changing climate—whether to a climate shock or to changing averages—depends heavily on the current state of the system it impacts, whether human, physical, or ecological. Thus, a small drought may be manageable for a farmer coming out of a prosperous year but ruinous if it follows another dry spell that exhausted the household's savings. Similarly, declining water runoff will be catastrophic for a region that already relies too much on its underground water resources, but it may be manageable for another whose agriculture is sustainable in current conditions.

Decades of environmental mismanagement have diminished ECA's natural resilience. Under the Soviet system economic growth was pursued in blatant disregard to natural conditions. When water was needed for irrigation, the rivers feeding the Aral Sea were diverted to the desert to produce rice, fruit, and cotton. Uzbekistan became one of the world's largest exporters of cotton, but at the cost of destroying the Aral Sea in the process. Today, the sand and salt blown from the dried-up sea bed onto the surfaces of Central Asian glaciers are accelerating the heat-induced melting of the glaciers—the source of most of the region's water. Uzbekistan's agriculture and hugely wasteful irrigation system are extremely vulnerable to climate change. The environmental legacy of central planning is particularly dramatic for agriculture and greatly increases the sector's vulnerability to climate change. Uzbekistan is not the only country to have specialized in producing a small number of crops ill-suited to the local environment; other countries and sub-national regions have as well. Poor management of soil erosion, water resources, pest control, and nutrient conservation makes the agricultural system especially vulnerable.

Over the next couple of decades non-climatic factors, such as legacy issues and continuing unsustainable demand, will be the main drivers of water stress in Eastern Europe and Central Asia (Vörösmarty et al. 2000). Similarly, floods cannot be explained by increased precipitation alone but result from a combination of heavy precipitation and poor land use and river basin management. Overall climate-related changes to freshwater systems have been small compared to factors such as pollution, inappropriate regulation of river flows, wetland drainage, reduction in stream flow, and lowering of the groundwater table (mostly due to extraction for irrigation). Clearly, more sustainable practices will be needed over the next decade before global warming's impacts become more severe.

Pollution is another legacy issue that magnifies the impact of climate change. While Estonia's coast is not generally vulnerable to sea-level rise, one danger persists: the leaching of radioactive waste at the Sillamae industrial center is separated from the sea by a narrow dam that is threatened by coastal surge. Coastal landfills around the Black Sea, notably in Georgia, have been identified as pollution hotspots, and coastal erosion could increase the amount of pollutants flushed into the sea, threatening a fishing industry already struggling with the consequences of overfishing and pollution.

In many parts of ECA, dangerous facilities or dump sites were often located close to weather-sensitive sites or heavily settled areas. This means that floods or extreme events can cause far greater damage here than would be the case in other parts of the world. Poor-quality housing will raise the human toll of climate change as heat waves turn poorly ventilated buildings into furnaces, and heavy rains brings leaks and mold. This is a special problem for ECA's cities—most of which have a glut of aging Soviet-era buildings made with prefabricated concrete panels and in desperate need of refurbishment.

Meanwhile, during the transition from central planning, ECA's abundant and over-dimensioned infrastructure has suffered from years of under-investment. Poor management often compounds the situation—especially in water and sanitation utilities. Global warming has had an especially negative effect on water systems—exacerbated by the inefficiency of systems of most water utilities, which under-price and suffer severe physical losses. This translates into high consumption and limited funding for upgrades and investments.

Elsewhere across ECA, the power sector is hard pressed to respond to the peaks in electricity demand linked to rising summer temperatures and is badly in need of upgrade and expansion. Warmer summers, with periods of intense heat, have strained the transmission networks of Azerbaijan, Kazakhstan, and Turkey, as well as systems throughout Southeastern Europe. In addition, extreme weather threatens the ability of networks to function as intended—especially aging and poorly maintained facilities.

ECA's transport infrastructure, with poorly maintained roads and structures, is also at risk. More intense precipitation will make subgrade pavement less stable and weaken retaining walls. Long periods of drought can lead to settling of the earth beneath the structures. More extreme temperatures will add to road deterioration as has already happened in Kazakhstan, where truck travel has to be limited on hot summer days when the asphalt softens.

It is tempting—though incorrect—to expect growth and prosperity to increase resilience to climate change. This is especially the case in Eastern Europe and Central Asia, where growth has typically occurred at the expense of the environment, thereby increasing vulnerability. In fact, growth and economic development are in some cases exacerbating vulnerability—such as coastal developments around the Black Sea, where buildings are being erected on sites exposed to coastal surge and storms.

Even Countries and Sectors that Could Benefit from Climate Change Are Poorly Positioned To Do So

Higher latitudes could benefit from improved conditions for agriculture: the Baltics, parts of Kazakhstan and Ukraine, and most of Russia (except for the North Caucasus). However, the potential for gain is unclear since it could be offset by increased variability and extreme events. Most countries will face a mix of losses and gains.

Nevertheless, many global studies about future food production assume ECA countries will help offset the decline in world food production resulting from decreasing yields in lower latitudes. In particular, Kazakhstan, Russia, and Ukraine are often mentioned as the countries with the world's greatest unrealized food production potential. The fact is that the current gap between potential and actual yields in these countries is significantly higher than any potential gains from climate change. In particular, the current yield gap for the former Soviet countries in Europe (including Ukraine and European Russia) is 4.5 times higher than the potential increase in productivity from climate change by 2050 (Olesen and Bindi 2002). In other words, unless current inefficiencies are addressed, the world's greatest unrealized food production potential will remain unrealized.

Forests show a similar pattern to agriculture. Estimates indicate that the largest share of potential forest stock increases in Europe would be from improved management (60 percent to 80 percent) rather than climate change (10 percent to 30 percent) (Easterling et al. 2007). Improving management requires strong forest institutions, which are often lacking in the transition countries.

The inability of Kazakhstan, Russia, and Ukraine to close the productivity gap or respond to recent crop price increases does not bode well for their capacity to adapt to and benefit from climate change. Indeed, the key challenge will be to close the existing productivity gap rather than ride the climate change wave to a new time of prosperity. That will depend on technology, policy, investment, support services, and crop management—and not simply on climate conditions. Northern areas will see intense competition between forestry and agriculture for land. The relative feasibility of field crops, tree crops, and livestock may further alter land-use patterns. A program of increasing farm outputs by expanding cultivation into newly temperate lands would require large investments in land clearing, production, marketing, and transport infrastructure—suggesting that improving the productivity of land currently under cultivation is more attractive.

The Next Decade Offers a Window of Opportunity for ECA Countries

Much of the adaptation needed to make ECA more resistant to climate change will have substantial co-benefits. Improved water resource management, better performing water utilities and energy systems, and upgraded housing and transport infrastructure are crucially needed independent of climate change. The potential gains from improved agricultural practices are much more significant than the benefits expected from climate change. And regardless of climate change, the region must clean up environmental hotspots, accelerate disaster management, and expand hydromet services.

Climate change exposes ECA's weaknesses while exacerbating their costs and risk implications. But where to start? Consistent with the advice of many experts on climate change adaptation, ECA should focus on areas and sectors already vulnerable to today's climate conditions and on actions that have immediate positive impacts for the population. In fact, much of what is discussed in this book falls into the category of "no-regret" actions—actions that are beneficial, whatever the climate change scenario.

But some decisions about long-term investments have to be made now—under conditions of uncertainty. For example, Albania, which currently derives 97 percent of its electricity from hydroelectric plants but cannot rely on it as a future source, must think through its long-term electricity strategy. Central European countries such as Poland, with over 5 million flats in poor Communist-era buildings, need renovation plans given the predicted increases in both rainfall and temperatures.

Uncertainty can be paralyzing. It is one of the reasons that a high *potential* for adaptation does not guarantee adaptation action. A recent study of the United States—often assumed to have a high capacity for adaptation given its wealth, technical resources, and large size (which allows for both diversification and spreading of climate risk)—shows that many at-risk organizations and individuals are failing to adapt (Repetto 2008). The Army Corps of Engineers is rebuilding Louisiana's levees to the same standards that failed during Katrina; many states in the arid Southwest are failing to incorporate climate change in their drought preparedness plans. In most cases, the reason for not changing standards or continuing to build in the same exposed location is uncertainty about "what to adapt to."

However, some countries and communities are not waiting. Australia and the United Kingdom have developed methodologies, standards, and databases to help organizations and individuals develop adaptation plans (UKCIP 2003, Australian Government 2005). One approach gaining traction is to focus on "robust strategies"—strategies that are effective in the face of an unpredictable future (Lempert and Schlesinger 2000). This approach tries to answer the question: *What actions should we take, given that we cannot predict the future?* It views climate change policy more as a contingency problem (*What if?*) than an optimization problem (*What is the best strategy given the most likely outcome?*). Looking for strategies that are robust to a variety of climate—rather than optimally adapted to the climate of the past—is essentially scenario-based planning and can help overcome the paralysis associated with uncertainty.

Perhaps the most critical lesson on how to develop adaptation plans is the importance of involving stakeholders. Stakeholders understand current vulnerabilities, which are the starting point for identifying future adaptation needs, and often have good ideas on how to reduce them. Involving stakeholders also improves the chance that the adaptation plan is implemented and that adaptation concerns are mainstreamed. This lesson becomes evident from the case in London where, five years after the *London's Warming* report, original stakeholders are still involved in the city's adaptation strategy.

ECA countries need to act. They can learn from other countries on how to manage uncertainty and assemble the right information to guide climate-resilient practices. But in ECA, perhaps more than in any other region, uncertainty should be a catalyst for action instead of an excuse for inaction. Fixing the region's current weaknesses and tackling its dismal environmental legacy will have immediate and substantial benefits for the welfare of individuals and for future economic growth, regardless of climate change.

Notes

1. The ECA region covers all the countries in the former Soviet Union, the transition countries of Central and Southeastern Europe (excluding East Germany), plus Turkey.

CHAPTER 1

A Framework for Developing Adaptation Plans

Marianne Fay and Jane Ebinger

Temperature increases of 1.6°C to 2.6°C are expected in Eastern Europe and Central Asia (ECA) by 2050 (chapter 2) even with global efforts to reduce greenhouse gas emissions. In the absence of such efforts, under a business-as-usual scenario, much greater warming is likely, with a median global warming of 5°C by 2100 (Sokolov et al. 2009). Such warming would be unlike anything the world has seen for more than 800,000 years. In fact, the difference between our world and the last ice age is only 5°C.

The implications for ecological and human systems are serious, even with relatively low 2°C temperature increases (Smith et al. 2009). Moreover, tremendous lags and inertia in the climate system imply that the climate will keep changing and that the sea level will continue rising for centuries as a result of carbon dioxide (CO_2) concentrations *currently* in the atmosphere (Solomon et al. 2009). In these circumstances, adaptation is not merely optional.

Still, the question remains: *What exactly should countries be adapting to?* Certainty about global trajectories coexists with substantial uncertainty about discrete events or local changes—particularly concerning precipitation. This uncertainty increases with attempts to downscale on a regional or subregional basis (chapter 2) and is amplified by the deterioration in national climate services in a number of Eastern European and Central Asian countries. The possibility of *reducing* the uncertainty about the nature and impact of climate change is limited—particularly in the short run. Policy work on climate change must therefore strive to help decision makers *manage* uncertainty, even as the scientific community tries to reduce it.[1]

Of course, managing uncertainty is what most business and policy makers do on a regular basis.[2] Managing uncertainty in the adaptation context means reducing the *vulnerability* of systems—social, economic, and ecological—to climate change—that is, reducing "the degree to which a system is susceptible to, and unable to cope with, adverse effects of climate change, including climate variability and extremes" (IPCC 2007a: 883). Reducing vulnerability begins with understanding its sources.

Vulnerability can be gauged without having to undertake expensive in-depth assessments. Countries and cities around the world have developed frameworks to assess and manage climate risk or reduce vulnerability (Australian Government 2005; Finnish Environment Institute 2007; Gagnon-Lebrun and Agrawala 2006; Ligeti, Penney, and Wieditz 2007; Natural Resources Canada 2005). These offer practical methodological approaches and lessons regarding implementation.[3]

But developing and implementing adaptation strategies entail a number of challenges.[4] It requires getting the right data and knowing how to use it. Climate data are usually projections based on large-scale models that lose reliability even as they are downscaled (chapter 2). Even where data are available, decision makers may be swamped by the number and variety of projections by climate modelers, or they may not understand how to use the data and manage the uncertainty in projections. Another challenge is to appraise and choose among options. Decision makers need to decide whether and when to undertake costly and irreversible investments in a situation where the magnitude and probability of risk are largely unknown. How should planners decide whether to spend millions to protect against a flood that may never come? When should farmers decide to switch crops or invest in irrigation? A critical element of any adaptation strategy is therefore a methodology for decision making under uncertainty.

In addition, obstacles to adaptation must be understood. These may be straightforward, such as technical or financial obstacles, or less obvious, such as those linked to human psychology or informational and cognitive barriers. For example, people already dealing with multiple uncertainties may have a "finite pool of worry" (Hansen, Marx, and Weber 2004). People may not act upon good information, like Floridian homeowners who fail to invest a few hundred dollars in simple upgrades that would greatly reduce their homes' vulnerability to hurricanes (Lewis 2007).

This chapter discusses vulnerability and its sources, as well as a way to estimate the vulnerability of countries in Eastern Europe and Central Asia. It then reviews how adaptation plans can and have been developed, including ingredients for success. The chapter concludes with a discussion of the challenges of making adaptation effective.

Vulnerability as a Function of Exposure, Sensitivity, and Adaptive Capacity

Vulnerability is the degree to which a system is likely to experience harm due to exposure to a hazard (Turner et al. 2003). Disentangling the components of vulnerability has been the subject of vigorous academic debate.[5] However, there is a simple and widely accepted approach that is broad enough to capture the essence of the different concepts of vulnerability detailed in the literature. The framework defines vulnerability as a function of exposure, sensitivity, and adaptive or coping capacity (figure 1.1).[6] The advantage of this approach is that it helps distinguish among what is exogenous, what is the result of past decisions, and what is amenable to policy action.

Exposure is a fairly straightforward concept: it is determined by the type, magnitude, timing, and speed of climate events and variation to which a system is exposed (for example, changing onset of the rainy season, higher minimum winter temperatures, floods, storms, and heat waves).

But the impact of a climate shock or change also depends on how *sensitive* a system is to that shock. The impact of a flood, for example,

FIGURE 1.1
Conceptual Framework for Defining Vulnerability

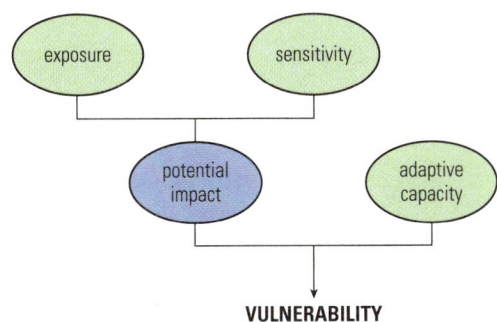

Source: Australian Government 2005.

will be the result of several factors: Do people live in the flood plain? Have toxic waste or water treatment plants been sited in the flood plain? Does the municipality have the organizational and financial resources to prevent the spread of waterborne diseases, help people access shelter, and quickly rebuild washed-out infrastructure, thereby reducing postdisaster loss of life and promoting faster recovery?

Sensitivity depends on how stressed the current system is. A system or a population already close to its limits will suffer great damages even from small shocks. Examples include poor individuals without any savings, congested and poorly maintained transport systems, populations in poor health, or water basins depleted of underground water resources.

Together, exposure and sensitivity determine the *potential impacts* confronting a community or a system—the impacts without considering adaptation. However, vulnerability also depends on how capable a system is of adapting and coping. Adaptation can be planned or autonomous; it can be anticipatory or reactive. The ability to adapt is a function of organizational skills, access to and ability to use information, and access to financing.

The distinction between sensitivity and adaptive capacity can be blurry. Sensitivity can be the degree to which a system is affected (positively or negatively) *in its current form* by a climate trend, climate variability, or a climate shock. However, adaptive capacity is dynamic and affects future sensitivity. In practice, the same factors that determine current sensitivity may also determine the extent of adaptive capacity. A poor household will be sensitive to shocks; it will also usually have less adaptive capacity due to its lack of resources to finance relocation or protective infrastructure such as dykes, stilts, or irrigation systems.

The exposure-sensitivity-adaptive capacity approach helps to identify the combination of factors that amplify or reduce the impact of climate change and to distinguish exogenous factors (exposure) from those amenable to local policy actions (adaptive capacity—hence, future sensitivity). It can be applied to particular regions or cities or sector by sector, as illustrated by the Australian government's application of this framework to agriculture (table 1.1).

A Vulnerability Index for Eastern Europe and Central Asia

We applied our vulnerability approach in an attempt to develop a simple vulnerability index for ECA countries. This is only a quick summary offered to guide more in-depth questioning. In particular, conditions within countries may vary substantially. Fay and Patel

TABLE 1.1

Applying the Vulnerability Framework to the Australian Cropping Industry

Vulnerability criterion	Findings
Exposure	The Australian broadacre cropping industry is located across a wide range of agro-ecological zones that differ significantly in their access to rain-fed and irrigated water. Climate change will leave many regions exposed due to increased temperature, reduced annual rainfall, or reduced water when needed for plant growth. An increase in the intensity and frequency of extreme events, such as drought and hail, will limit the capacity to grow productive crops in some regions. It is likely that the areas now considered marginal in their capacity to produce viable crops will be the most vulnerable to climate change.
Sensitivity	Some regional locations will be more sensitive to the impacts of climate change than others. This sensitivity can be attributed to a number of factors, including heat stress, susceptibility to pests and diseases, seasonal rainfall patterns delivering rain when it is not needed, frequency of frost days and very hot days, number of drought years, and the ability to recover after drought.
	Some regional communities depend heavily on the economic viability of the broadacre cropping sector and so will likely suffer significant decline subject to the impacts of climate change. The impact of elevated CO_2 on plant growth, together with reduced rainfall and increased temperature, may provide opportunities in some regions; but it is more likely to increase the reliance on nitrogen fertilizer to achieve current production rates.
Adaptive capacity	The broadacre cropping industry has few options to adapt to the impacts of climate change and relies strongly on the ability to obtain a good return in one year out of three. The introduction of drought tolerance into new plant varieties will increase the adaptive capacity of agriculture within a limited range of increased temperature and reduced soil moisture conditions. In addition, improved water-use efficiency through better soil management, such as no-till, will increase the capacity of the industry to adapt to small changes in climate.
Adverse implications	Broadacre cropping industries remain the lifeblood of regional Australia, with crop production worth about AU$8 billion of export and domestic earnings annually (principally export). Any adverse impacts of climate change will have a significant detrimental impact on regional communities.
Potential to benefit	The broadacre cropping industry has limited opportunity to adapt to the impacts of climate change, with the main adaptations likely to be short to medium term only. Adaptations that will provide ongoing productivity in some regions (although the most marginal regions will be the most exposed) are increased water-use efficiency through soil management, increased drought tolerance and shorter season varieties, long-term weather forecasts, a move away from commodity trading, and improved protection from pests and diseases. In all cases, the potential to benefit improves where more accurate and reliable annual forecasts are provided, improving the capacity to increase yield and reduce the number of failures in bad years.

Source: Australian Government 2005.

(2008) discuss in detail the methodology and underlying data sources used in developing this index.

Our vulnerability index combines three subindices, capturing a country's exposure, sensitivity, and adaptive capacity.[7] The first, *exposure,* is based on an index measuring the strength of future climate change relative to today's natural variability (Baettig, Wild, and Imboden 2007). The index is available on a country basis and includes both annual and seasonal temperature and precipitation indicators. It combines the number of additional hot, dry, and wet years; hot, dry, and wet summers; and hot, dry, and wet winters projected over the 2070–2100 period relative to the 1961–1990 period. The index suggests that the countries most exposed to future climatic change are the Russian Federation, Albania, Turkey, Armenia, and, to a lesser

FIGURE 1.2
An Index of *Exposure* to Climate Change

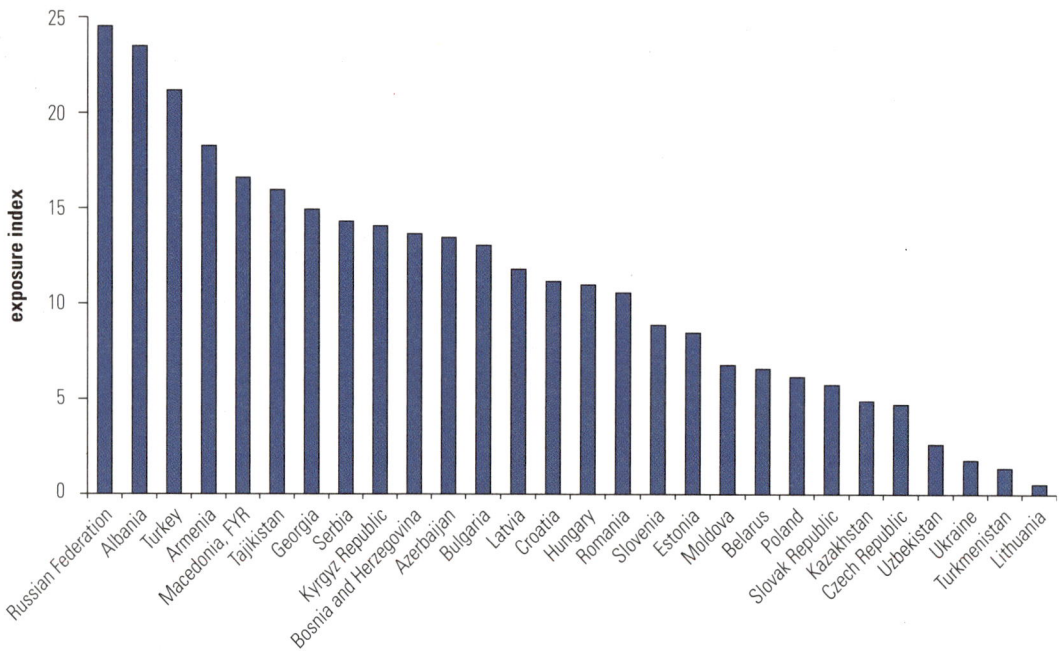

Source: Baettig, Wild, and Imboden 2007.

extent, the Former Yugoslav Republic of Macedonia and Tajikistan (figure 1.2).

The second subindex, a country's *sensitivity* to climate change, is based on indicators likely to increase the impact of climate shocks. These include physical indicators, such as the available renewable water resources per capita and the extent of air pollution (since particulate matter in the air worsens the impact of heat waves), and economic indicators capturing the importance of agriculture in the economy (share of employment and value of assets) and the share of electricity derived from hydroelectric plants. We also included a measure of the overall quality of infrastructure since infrastructure in poor condition is more likely to fail during an extreme event. Last, we included the share of population over age 65, since people in this group tend to be more sensitive to climate shocks. The results suggest that Central Asian countries are particularly sensitive to climate change—along with Albania, Armenia, and Georgia (figure 1.3).

The third subindex, *adaptive capacity,* is estimated by combining social (income inequality), economic (gross domestic product [GDP] per capita), and institutional measures.[8] The adaptive capacity index differs from the other two in that higher values are good—that is, they denote higher adaptive capacity (figure 1.4). As expected, we

FIGURE 1.3
An Index of *Sensitivity* to Climate Change

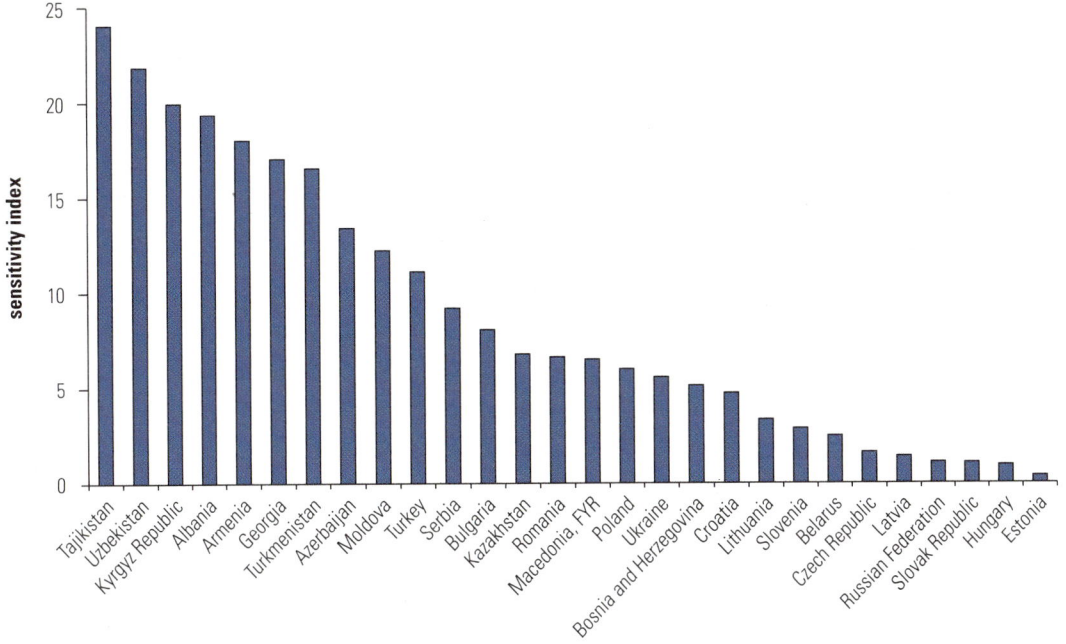

Source: Fay and Patel 2008.

FIGURE 1.4
An Index of *Adaptive Capacity* to Climate Change .

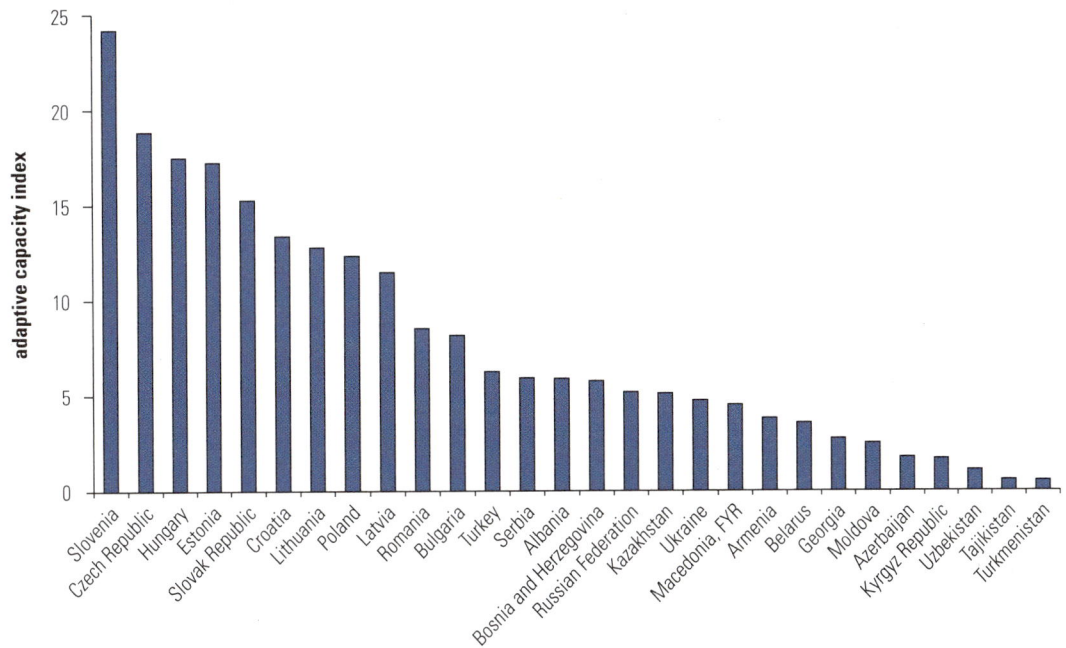

Source: Fay and Patel 2008.

find sensitivity and adaptive capacity to be inversely correlated, with the richer countries generally less sensitive and with higher adaptive capacity. This finding is not universal, however: the Russian Federation has both low sensitivity and low adaptive capacity.

Combining the three components into a single index of vulnerability yields the ranking shown in figure 1.5a. The figure uses a different scale to enable us to see which factor—exposure, sensitivity, or lack of adaptive capacity—drives countries' vulnerability. Thus, among the most vulnerable, Albania suffers from relatively high exposure, while the Kyrgyz Republic and Tajikistan are estimated to have social and productive structures that make them very sensitive to the impact of a changing climate. The Russian Federation stands out for its high exposure and limited adaptive capacity, which are offset by relatively low sensitivity. Figure 1.5b rescales the overall index to give a simple indicator of vulnerability.

An Alternative Measure

A good alternative proxy for vulnerability to climate change is the current "adaptation deficit"—the vulnerability to the current climate. But even this can be hard to estimate on an aggregate basis.

One proxy is offered by the incidence and impact of natural disasters over the most recent decades (figure 1.6). This suggests that the most vulnerable countries in Eastern Europe and Central Asia are Albania, Tajikistan, and Moldova. The ranking corresponds somewhat to that suggested by our index of vulnerability: the two most vulnerable countries are the same (Albania and Tajikistan), but the overall correlation of the two sets of ranking is only about 0.5.

Disaster impact data,[9] however, have many limitations as a measure of the deficit of needed adaptation to climate change. First, they look only at damage from natural disasters, but climate change also entails increasing variability and shifting averages, which require different adaptation responses. The ability to limit disaster damage is not necessarily the same as the ability to handle changes in variability and averages, and vice versa.

Second, financial estimates of damages are considered much less reliable than the data on number of deaths; but deaths are a flawed indicator because adaptation to climate change is about much more than saving lives. The deficit measured by deaths is the inability of states to protect human life in disaster conditions—not the inability to protect livelihoods, welfare, and economic production in conditions of changing means and variability as well as disasters.

FIGURE 1.5
An Index of *Vulnerability* to Climate Change

a. The Drivers of Vulnerability to Climate Change

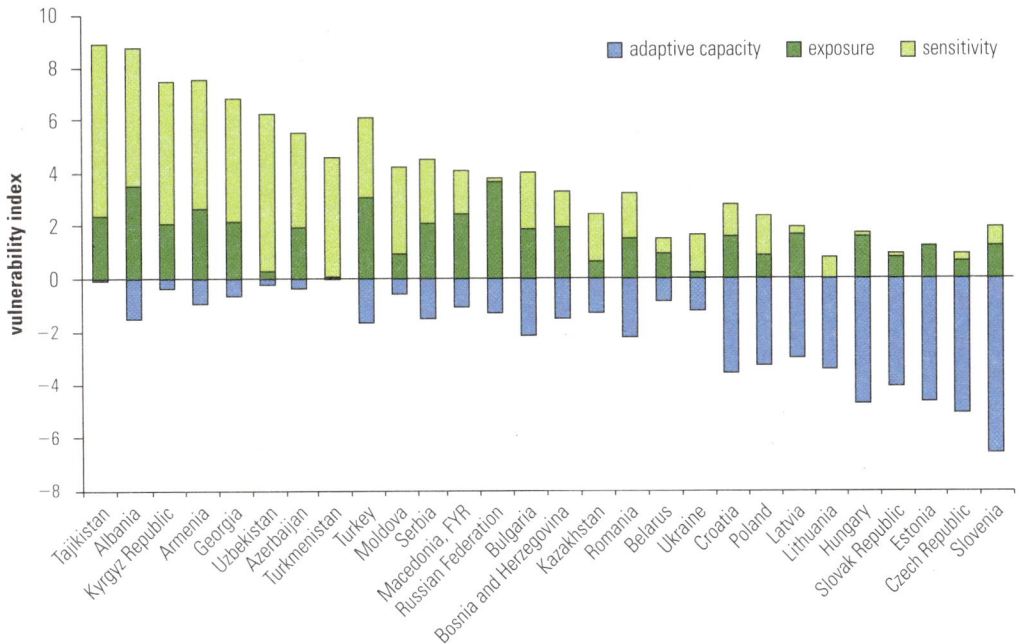

b. A Simplified Index of Vulnerability to Climate Change

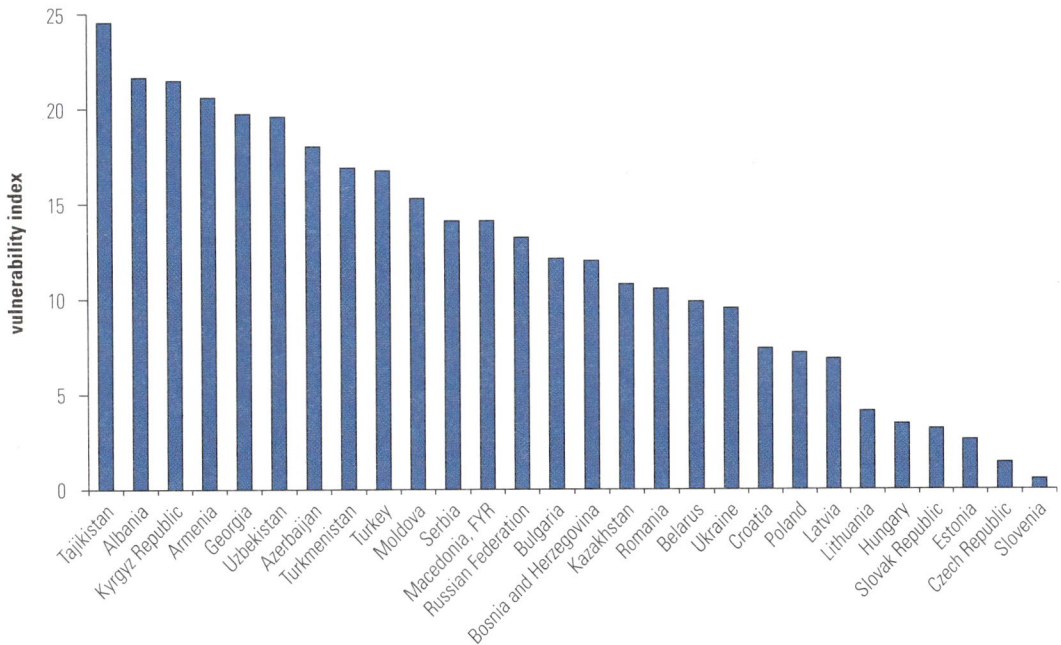

Source: Fay and Patel 2008.

Note: Adaptive capacity decreases vulnerability, and hence, it is shown in (a) as taking negative values. Slovenia has very high adaptive capacity, which is therefore large and negative, while Tajikistan has very low adaptive capacity, which is therefore close to zero. In (b), the overall indicator is rebased to vary from zero to 25, to be comparable to figures 1.2, 1.3, and 1.4.

FIGURE 1.6

Impact of Natural Disasters in Eastern Europe and Central Asia, 1990–2008

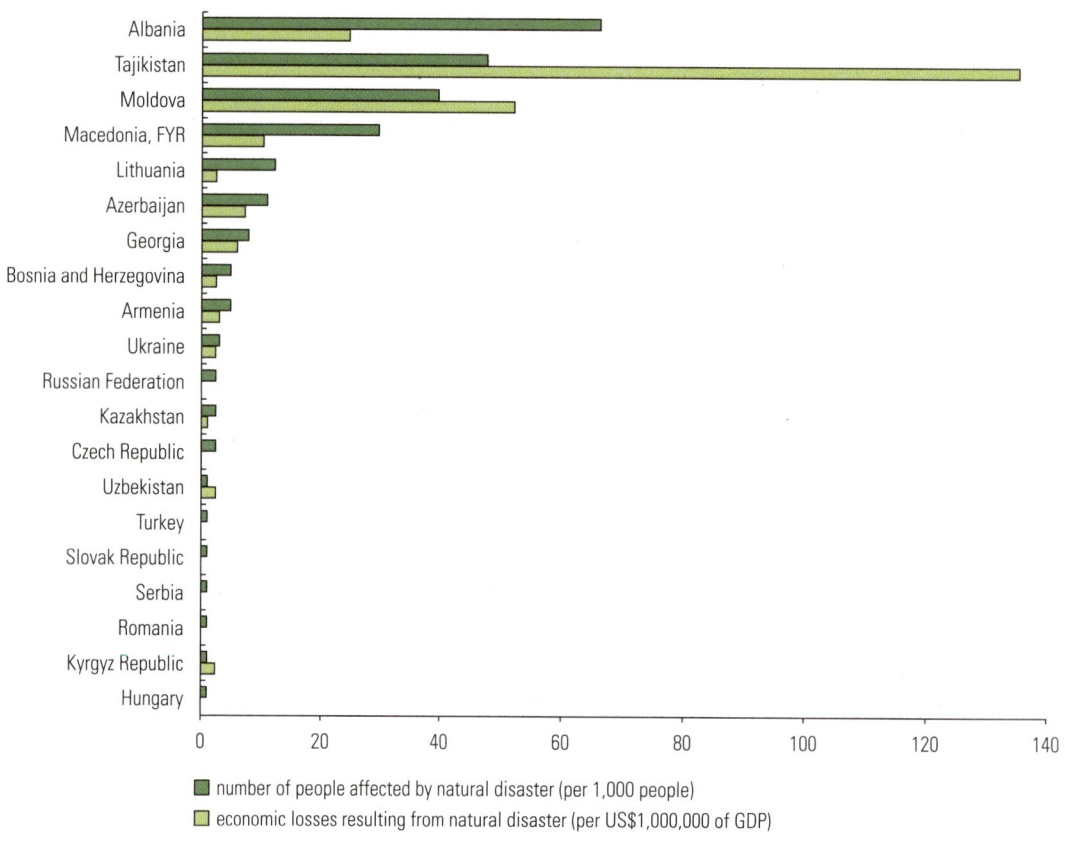

■ number of people affected by natural disaster (per 1,000 people)
□ economic losses resulting from natural disaster (per US$1,000,000 of GDP)

Source: EM-DAT 2008.

From Vulnerability to Action: Tackling the Challenge of Adaptation

Awareness and concern about climate change are relatively low in Eastern Europe and Central Asia. Only about 50 percent of those interviewed in the 7 ECA countries included in a recent Global Attitude Survey of 48 countries consider climate change to be a serious problem—lower than the 59 percent average across the full sample of 48 countries (figure 1.7).

Part of this problem stems from the often confusing or misleading way in which climate change is communicated to the public. Journalists tend to want to present a "balanced" story, which has resulted in the media granting equal weight to climate scientists and to commentators who often lack academic credentials.[10] Scientists fault this traditional application of balance in reporting science, arguing that

FIGURE 1.7

Global Warming: How Serious a Problem?

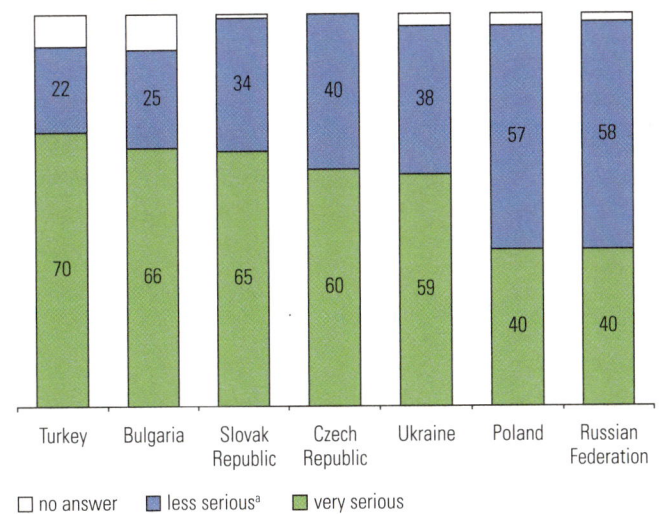

| | no answer | | less serious[a] | | very serious |

Source: Pew Global Attitudes Project 2007.
a. Less serious = somewhat serious, not too serious, or not a problem.

peer-reviewed studies should not be weighed equally against expressions of opinion or policy arguments. This has led to calls for improved communicaiton between journalists and scientists.

In addition, *awareness* or understanding of climate change does not necessarily lead to action and behavioral change—as shown by ample evidence from risk, cognitive, and behavioral psychology. A recent review of adaptation action in the United States shows continued, chronic failure to upgrade standards to more appropriate ones despite the existence of information on increased likelihood of floods, droughts, and hurricanes (Repetto 2008). Part of the issue is that individuals may have a finite pool of worry and so may fail to act upon existing information because of competing priorities or a lack of experience with climate-related events. They may feel unsure about how to change their own behaviors to address such a large-scale problem, or they may face insurmountable financial or technical obstacles.

Policy makers need to be aware of these cognitive, behavioral, and sociocultural barriers to action. While consistent and simple information campaigns are important, they will not guarantee that action will follow. Policy makers must be ready to address perceptions of risk and vulnerability, in addition to capacity to adapt. Useful tools to support adaptation planning are discussed in the following sections.

Approaches to Adaptation Planning

Hazard identification and risk assessment require a comprehensive approach, even if the analysis is only qualitative. Vulnerability should be assessed regardless of the uncertainities or lack of data about future climate and socioeconomic conditions. The initial assessment should be based on the *current climate,* identifying current vulnerabilities and knowledge or other gaps. Such an assessment should include feedback loops throughout the process so that decision makers can screen, evaluate, and prioritize risks before deciding on more detailed analysis of certain measures.

Most assessment frameworks emphasize a number of steps common to hazard identification and risk assessment (Australian Government 2005; EEA 2007; UKCIP 2003). Box 1.1 offers more detail on these, but the key steps progress as follows:

- Identify hazards and assess risks.

- Identify options to manage risks.

- Appraise options, including their costs and benefits.

- Make a decision.

- Implement the decision.

- Monitor, evaluate, and review periodically.

This approach offers flexibility; it can be applied at different levels, from national ministries down to a city government or municipal water utility. Its usefulness at the local level is key since adaptation actions are often locally determined and implemented. However, a mayor or utility manager will more likely succeed in gathering the needed guidance, informational inputs, and political momentum if a national assessment has been conducted and a national adaptation plan, linked to local efforts, is in place (Adger et al. 2007).

The approach's low cost presents another advantage. While adaptation measures may ultimately be costly, the processes of screening risks and developing an effective adaptation plan are within the capacity of government budgets. The United Kingdom Climate Impacts Programme (UKCIP), the agency in charge of helping promote adaptation across the country, has only 10 full-time staff members and a budget for three years (2002–05) of about US$3 million. This is possible because sector stakeholders, not permanent staff members, are the real engines of activity.

Stakeholder engagement may be the most important lesson emerging from a review of adaptation processes.[11] This is critical for

BOX 1.1

Standard Approaches for Understanding Risk and Developing an Adaptation Strategy

Conducting a qualitative risk assessment includes the following key steps:

1. **Establish the context and objectives for the assessment.** Formulate the issue and the scope of assessment; define the objectives of the exercise and the broad context for the decision; identify climate scenarios; and define the geographic region and key stakeholders (government, sector, and community) or audience to whom the assessment is targeted.

2. **Identify the hazards.** Start with a screening exercise to identify the main hazards, including what could happen under different climate scenarios. Structured brainstorming techniques involving key stakeholders (for example, policy makers and experienced sector specialists), such as the "Structured What If Technique" (SWIFT), can help identify major hazards. SWIFT screens hazards by considering deviations from business as usual or normal operations, using checklists to support a brainstorming exercise. This technique allows a systematic, high-level, team-oriented approach but relies on the quality of the expert team. For more details, see http://rmd.anglia.ac.uk/uploads/docs/SWIFT.doc or HSE (2001).

3. **Analyze the hazards.** Consider each major hazard in step 2 and existing safeguards or controls, including policy and management responses. Assess the consequences to the system on the basis of existing controls and make a judgment about the likelihood of those consequences materializing. Determine the level of risk. Australian Government (2006); HSE (2001, 2006); New Zealand Climate Change Office (2004); and UKCIP (2003) provide good examples of risk matrices and their application. (See below for a further discussion of decision making under uncertainty.)

4. **Evaluate the risks.** Rank the risks, screening out minor risks and prioritizing major risks for further analysis. Describe the uncertainties associated with each risk factor and the sensitivity of the analysis to assumptions.

5. **Identify and appraise options to manage risk.** One methodology to support practical decision making is to identify a set of climate conditions as benchmark levels of climate risk that represent the threshold between tolerable and intolerable risk—which decision makers should work to avoid. For example, for a hydropower system already under stress, a certain frequency of drought conditions could exceed the system's adaptive capacity and require proactive adaptation to address energy supply and security needs. (See table 1.2 for a typology of adaptation options.) Other methodologies include real options analysis (discussed in the main text) or insurers' approach to risk valuation. See HSE (2001, 2006) and UKCIP (2003) for more examples.

continued

BOX 1.1 *Continued*

6. **Develop an adaptation plan.** Develop a prioritized action plan that draws on the options identified to manage risk, including a review of costs and associated benefits, to adapt to identified vulnerabilities and risks. Discuss risks associated with adaptation itself: under-, over-, or maladaptation. Ensure adaptation planning responds not only to changing climate averages but also to increased variability and extremes.

7. **Implement the adaptation plan.** In formulating a roadmap for implementation, decide on whether to build on and update existing legal and regulatory frameworks, institutions, policies, strategies, and emergency and disaster management plans or to develop new arrangements altogether; determine what institutional capacity exists and what is needed to support implementation; assess financing needs and sources; and clarify what data and information gaps exist and how to address them, including through research and development.

8. **Review the action plan.** Establish monitoring and evaluation (a feedback loop) to periodically re-evaluate risks and priorities as more information becomes available or other relevant events occur.

Sources: Australian Government 2005, 2006; HSE 2001, 2006; UKCIP 2003; New Zealand Climate Change Office 2004; and http://rmd.anglia.ac.uk/uploads/docs/SWIFT.doc (accessed August 30, 2009).

multiple reasons: stakeholders such as local farmers, water engineers, utility managers, or public health staff members possess greater knowledge of stress points and vulnerability that may be difficult to access otherwise. They are also critical for making assessments and recommendations on the ground.

In addition, involving stakeholders in the planning process increases the chance that they will "own" and support the ensuing adaptation plan. Further, by involving stakeholders and local decision makers at all levels in an adaptation plan, governments (local or national) improve prospects that society will incorporate climate change concerns in future investment and management decisions (box 1.2).

The cities of Boston and London offer interesting contrasts in their approaches to adaptation strategies (Ligeti, Penney, and Wieditz 2007). In Boston, stakeholders were not consistently involved in the researcher-led process; as a result, interest waned so that the final study was poorly understood, was seen as overly technical, and had limited impact. However, London used a bottom-up approach involving stakeholders (local, regional, and national government representatives; utilities; business organizations; environmental nongovernmental organizations; and climate research staff) in the

BOX 1.2

Lessons on the Engagement of Stakeholders in Adaptation Plans— Urban Experiences

- Key stakeholders include municipal and regional government departments, transportation authorities, utilities, and conservation authorities.

- Engagement of key stakeholders is vital for understanding how climate change may impact cities, identifying practical adaptation strategies, and gaining support for implementing those strategies.

- Engagement of stakeholders often begins with an event designed to raise awareness and pique interest in climate impacts and adaptation. However, a plan for ongoing engagement of stakeholders after the event is also necessary.

- It is important to understand the general goals and concerns of stakeholders and to investigate the way in which climate change could affect these.

- Sign-off from senior management is important; however, engagement may be more successful with mid-level stakeholders, who will likely participate more consistently in the adaptation process and, therefore, develop a better understanding of impacts and adaptation strategies.

- Regular communications and meetings are required for sustained stakeholder engagement.

- Stakeholder engagement can be time consuming and costly; therefore, adequate staff time and funding are essential for successful and sustainable stakeholder involvement.

- Processes that focus on technical modeling issues and reports that contain too much technical jargon will reduce stakeholder engagement.

- Researcher-led adaptation initiatives are in danger of coming to an abrupt end when funding is over. For these initiatives to go beyond research to action, stakeholders must take ownership of the process

Source: Ligeti, Penney, and Wieditz 2007.

London Climate Change Partnership. After the *London's Warming* report was issued (Clarke et al. 2002), many continued working with the Climate Change Partnership, which evolved into a more permanent organization. Five years later, many of the stakeholder organizations are still involved and still participate in the steering group of the Climate Change Partnership—a complete contrast with the Boston experience (Ligeti, Penney, and Wieditz 2007).

Mainstreaming Adaptation into Development

An adaptation plan needs to be mainstreamed into the daily operations of private and public decision makers. Mainstreaming will ensure that climate stress is integrated with the multitude of other stresses that human and natural systems must cope with. It also helps avoid a portfolio of efforts that are unrealistic, inefficient, and potentially ineffective, particularly if they are working at cross-purposes (Yohe et al. 2007). However, experience with mainstreaming adaptation is short (Adger et al. 2007).

While most agree that adaptation efforts should build on existing activities rather than multiplying fragmented initiatives, two distinct approaches are emerging. The first takes a technological view, focusing on physical exposure and ensuring that climate variables and projections influence choices of technologies and infrastructure specifications. This approach has been optimistically called "climate-proofing." The second takes a development-oriented view, whereby projects are expanded to increase adaptive capacity and reduce vulnerability (Klein 2008). The latter approach raises the question of whether mainstreamed adaptation is just good development practice. Clearly, adapting to a changing climate makes development sense. Furthermore, as countries develop and get richer, they tend to accumulate the human, physical, and financial capital critical for adapting to changing conditions.

But adaptation is not just business-as-usual development. Development that fails to integrate climate considerations will not be sustainable; and in some cases, it may make populations or sectors *more* vulnerable to climate impacts (box 1.3). Examples include coastal developments that ignore sea-level rise; increased reliance on air-conditioning, without regard for efficiency and demand-side management; and investments in irrigation to maintain rural livelihoods no longer suited to a changing climate and hydrology.

Effective Adaptation Requires the Right Decision-Making Tools

Even as we agree that frameworks exist to develop adaptation plans and that lessons are being drawn as to how and why to involve stakeholders (box 1.2; Næss et al. 2006), technical challenges remain, especially if an adaptation plan is to change behaviors. Adaptation planning and implementation need to respond to both changing averages and increased variability and extremes.

BOX 1.3

Is Adaptation any Different from Development?

Haven't we been working to improve individual and institutional capacity all along? Haven't governments always known that agricultural extension services are important? Haven't engineers always had to handle uncertainties about rainfall and floods when building reservoirs?

These and other examples illustrate the many overlaps between good adaptation and good development, and between the institutional and human resources required for each. Klein (2008), among many others, advocates adaptation that is mainstreamed into development activities and not compartmentalized into strictly technical measures.

So how does adaptation differ from development? First, in concrete technological terms, parameter specifications and assumptions about climate are changing, so designs for infrastructure projects and assumptions about revenues from tourism or agriculture should change. Second, the amplification of uncertainty from unknown distributions of the incidence weather events requires decision makers to move away from business-as-usual development. Third, changing priorities may lead to different choices from among a set of investment or policy options. Finally, for better or worse, the international community is providing separate financing mechanisms, which may require differences in conceiving of and implementing projects.

Trade-offs, and even conflict, may arise in allocating resources for the different activities. For example, ceasing cultivation of irrigation-intensive cotton because of increasing water stress in Central Asia would cause significant income loss for today's farmers, at least in the near term. And economic development alone does not solve the adaptation problem: it does not remove enough people or enough natural and built resources from harm's way. Development strategies that do not incorporate adaptation priorities can exacerbate vulnerabilities.

Adaptation in the Context of Development

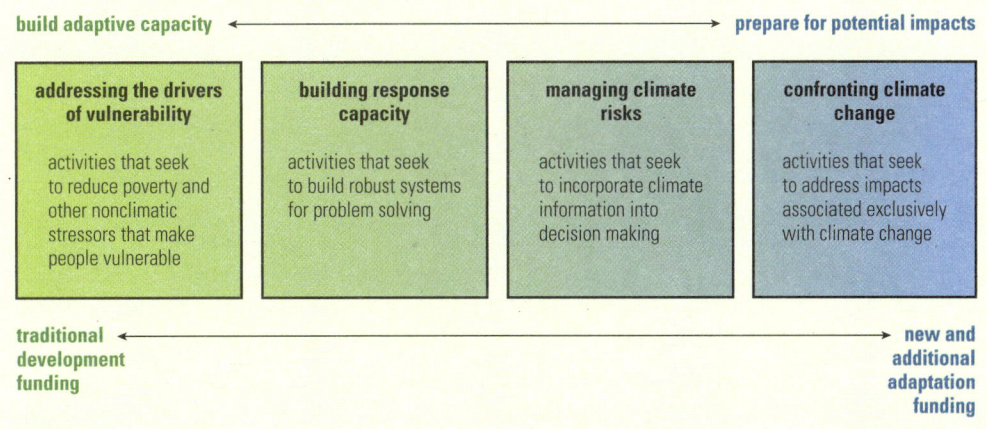

build adaptive capacity ←——————————————————→ prepare for potential impacts

addressing the drivers of vulnerability	building response capacity	managing climate risks	confronting climate change
activities that seek to reduce poverty and other nonclimatic stressors that make people vulnerable	activities that seek to build robust systems for problem solving	activities that seek to incorporate climate information into decision making	activities that seek to address impacts associated exclusively with climate change

traditional development funding ←——————————————————→ new and additional adaptation funding

Source: Adapted from Klein 2008.

Making Decisions under Uncertainty

Changing climate averages, increasing variability, and the rising frequency of extreme weather events—these are all reasons that decision makers in the private and public sectors must begin incorporating climate risk management in policy planning.[12] The question is how to do this given the inherent uncertainties in climate projections and the costs of either inaction or misguided adaptation.

Options to deal with or manage risk fall into several categories; even "do nothing" and "bear the loss" can constitute a strategy (table 1.2). A combination of different adaptation strategies may be optimal, as may be a shift from one to another as circumstances change. Whatever approach is taken, the literature offers some generic but important recommendations on how to select options for action: minimize irreversible investments and maximize reliance on win-win approaches, that is, approaches that yield benefits even if the projected risks do not materialize as expected.

TABLE 1.2
Typology of Possible Adaptation Strategies

Bear the loss	"Do nothing," where there is no capacity to respond, or the cost of adaptation is too high in relation to the risk or expected damage (such as loss of coastal areas or loss of a species).
Share the loss	Provide private insurance, public relief, reconstruction, and rehabilitation paid from public funds.
Modify the threat	Add flood control measures; promote migration of people from high-risk areas; encourage new agricultural crops; change location of new housing, of water intensive industry, and of tourism; improve forecasting systems to give advance warning of hazards and impacts; and develop contingency and disaster plans.
Prevent effects	*Structural and technological changes* needing more investment such as: provide increased irrigation water; supply increased reservoir capacity; encourage water transfers; endorse water efficiency; scale up coastal protection; upgrade waste water and storm water; build resilient housing; modify transport infrastructure; and create wildlife corridors. *Legislative, regulatory, and institutional changes* such as: change traditional land-use planning practices; provide more resources for estuarine and coastal flood defense; revise guidance for planners; include climate change risks in criteria for site designation for biodiversity protection; and amend design standards.
Change use	Where continuation of economic activity is impossible or extremely risky; for example, substitute for more drought tolerant crop or return crop land to pasture or forest.
Change location	Relocate major crops and farming regions away from areas of increased aridity and heat.
Research	Provide new technologies and methods of adaptation; improve short-term climate forecasting and hazard characterization; provide more information on frequency and magnitude of extreme events; offer better regional indicators for climate change; require more risk-based integrated climate change impact assessments; encourage better knowledge of relation between past and present climate variation and system performance; and produce higher resolution spatial and temporal data on future climate variability from model-based climate scenarios.
Educate, inform, and encourage behavioral change	Lengthen planning timeframes; reduce uneven awareness by stakeholders; and increase public awareness to encourage people to take individual action (health, home protection, and flood awareness) and to accept change to public policies (coastal protection, landscape protection, and biodiversity conservation).

Source: Adapted from UKCIP 2003 (table 2.3)

Many analysts recommend using past extreme events as an indicator of the range of risks to prepare for and of the key vulnerabilities of existing systems. They call for planners to develop vulnerability indices that rely on known dangers and hotspots (EEA 2005). This may be a good first step, based on the relatively strong certainty that the frequency of extreme events will increase. However, because climate projections suggest an increase in the *intensity,* not just frequency, of extreme events (such as worse floods, droughts, or storms), protecting against known dangers is not enough. And changing averages must be considered, which may alter whether something should be built or how something should be managed. For example: Will there be enough water for a hydroelectric power plant?

Decision makers will have to choose between competing options, with trade-offs between current costs and potential averted damages. Moreover, they must choose in a context of uncertainty about the probability and magnitude of the changes against which these options are to protect. Climate variability is nothing new, but uncertainty has increased. The probability distributions of extreme weather events are changing, and the extent and speed of this change are unknown. So-called 1,000-year floods (floods of such magnitude that they generally happen only once every 1,000 years) may now be 100-year floods. Such events have become unpredictable, as they no longer follow a known probabilistic pattern.[13] Decisions about long-lived infrastructure also must take changing averages into account.

Unknown probability implies that traditional cost-benefit analysis or maximum expected value approaches (such as minimax and maximin) cannot be used. The use of subjective probabilities (such as expert opinions) continues to be an object of research (see UKCIP 2003 for an overview). In addition, the possibility of major irreversible events or "unacceptable risks" may imply that cost-effectiveness analysis for a given level of risk is more relevant.[14] In other words, the best approach might be to determine what is the socially acceptable level of risk, and then to identify the most cost-effective measures to achieve this level. Multicriteria analysis—which complements techniques that rely on criteria expressed solely in monetary terms—can help to distinguish acceptable from unacceptable options (UKCIP 2003).

A focus on "robust strategies"—robust in the face of an unpredictable future (Lempert and Schlesinger 2000)—is well suited for dealing with unknown probabilities. This means answering the question: *What actions should we take given that we cannot predict the future?* as opposed to *What is the best strategy given that we expect a particular state of the world to occur?* Climate change policy then becomes a contingency problem rather than an optimization problem.

Looking for robust rather than optimal strategies amounts to scenario-based planning. This approach considers "shaping actions," which influence the future; "hedging actions," which reduce future vulnerability; and "signpost" events, which are chosen to trigger a change in strategy (van der Heijden 1996).[15] A recent documented case of a California water utility facing likely but uncertain decreases in water availability provides a practical example of how to apply this approach (Groves et al. 2008).

Getting the Right Data—and Knowing How To Use It

Policy makers do not have a climate equivalent to the kind of fiscal, financial, or public health data that are regularly published as a limited set of well-understood variables. One issue is that climate indicators relevant for a farmer are not necessarily the same as those relevant for an electrical utility manager or health specialist—and the data relevant for the corporate monoculture farm may be different from that relevant to the small orchard owner. It is difficult to define which data are needed, at what scale in time and space these data should be produced, and perhaps most important, how to interpret these data and cope with the inherent uncertainty. Nor is it always easy to access such climate projection data—particularly at a down-scaled level.[16] Stakeholder-based climate risk assessment and adaptation planning tools (outlined earlier in this chapter) can provide a basis for understanding climate data needs, gaps, and costs for a given sector, and can help to prioritize their importance.

Some practitioners of adaptation planning note that the need for climate projection data is overemphasized; such data are uncertain, and climate is only one driver of vulnerability. They also argue that an excessive emphasis on climate data can obscure the need for more important steps in adaptation planning, such as stakeholder involvement and understanding of vulnerability to current climate.[17] It is striking that most adaptation strategies reviewed for this book spend little time discussing projected climate change or tend to do so in broad terms—focusing on general trends (wetter or drier; hotter; more extremes) or highlighting major uncertainties.

There are now many initiatives to make climate data available, more accessible, and more useful for decision making. For example, the World Bank has developed a "climate portal" that allows users to obtain climate projections for particular locations based on the ensemble of models used by the Intergovernmental Panel on Climate Change as well as on a high-resolution Japanese General Circulation Model (box 1.4). For countries able to develop their own

BOX 1.4

The World Bank's Climate Portal

The Web-based World Bank Climate Change Data Portal provides quick and accessible global climate and climate-related data to the development community. The site is based on the familiar Google Maps platform and allows users to access outputs from climate models, historical climate observations, natural disaster data, crop yield projections, and socioeconomic data at any point on the globe. The site includes a mapping visualization tool (webGIS) that displays key climate variables and links to World Bank databases and a spatially referenced knowledge base. The portal will also serve as a launching point for climate change tools, and includes the ADAPT tool for assessing climate risk of World Bank projects.

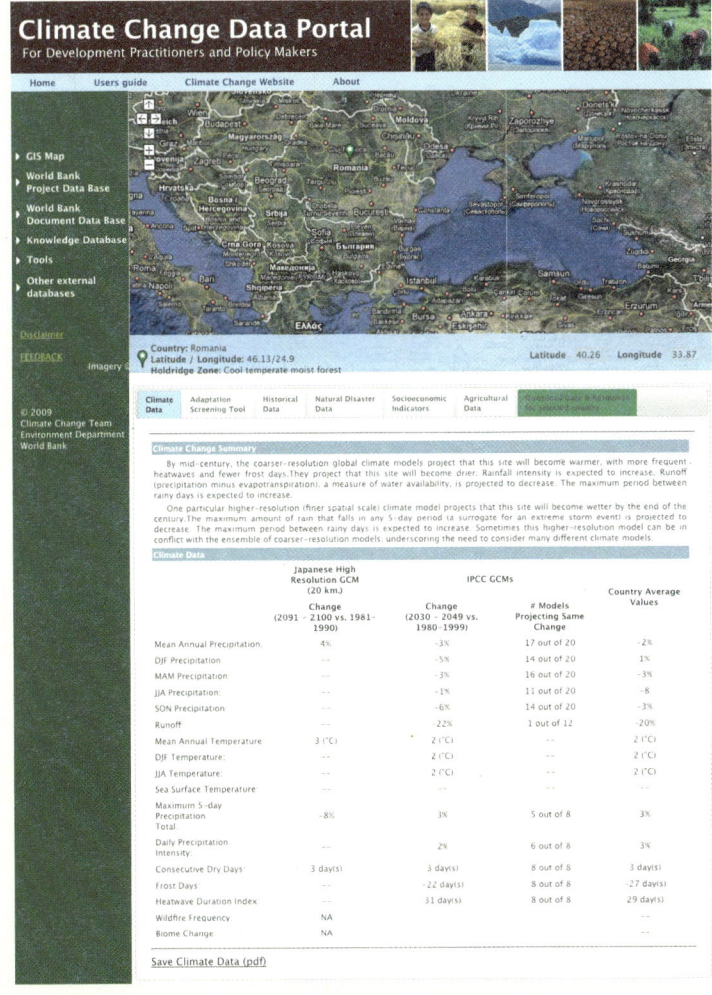

Source: http://worldbank.org/climateportal/.

BOX 1.5

Tools to Help You: Tools Portfolio of the UK Climate Impact Program

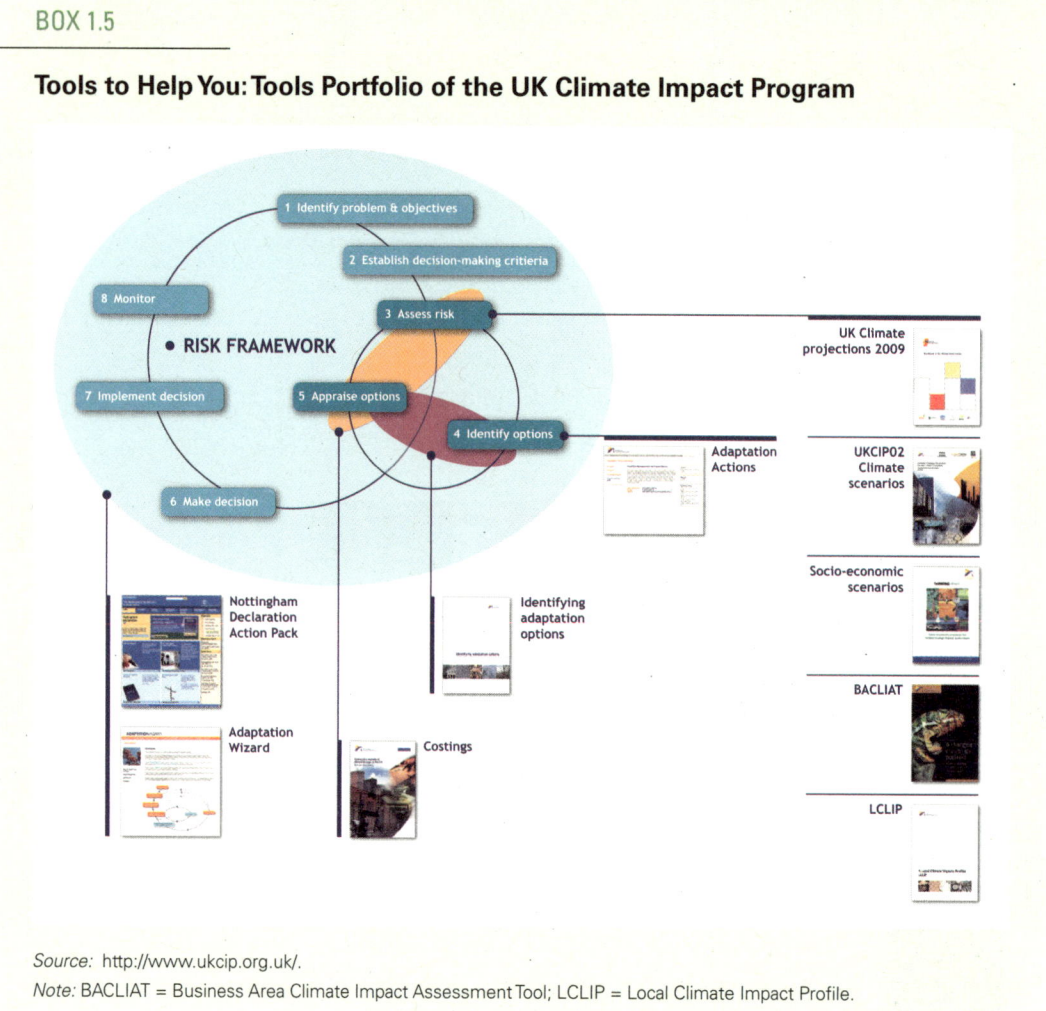

Source: http://www.ukcip.org.uk/.
Note: BACLIAT = Business Area Climate Impact Assessment Tool; LCLIP = Local Climate Impact Profile.

climate portals, the UKCIP Web site is a good example. It offers climate projections (termed climate scenarios on that site), as well as an "adaptation wizard" to help users work their way through potential vulnerabilities and methodologies to appraise options and make decisions (box 1.5).

Notes

1. See Schneider and Kuntz-Duriseti (2002) for a discussion of the need to manage, rather than master, uncertainty. They trace the origin of this approach to work on resilience in ecology.

2. Uncertainty, formally speaking, is different from risk (where there is a known probability distribution); however, it was impossible to write this book without using the word "risk," so we have not adhered to a formal use of the two terms.

3. These works stand in contrast to a large academic literature on adaptation that focuses on frameworks and definitions. While helpful in framing the discussion of what adaptation is, that literature does not offer much in the way of practical guidance for preparing and implementing an adaptation program.

4. Many strategies have been written up, and in some cases, attempts have been made to provide a critical analysis of what appears to work. For a comparative analysis of these strategies, see Heinz Center (2007).

5. A whole cottage industry has sprung up on how to define and measure vulnerability. This is partly because different disciplines or fields of research (for example, catastrophic risk management, ecology, social protection, and climate change) use similar terms for different purposes, or use many different terms to describe the same fundamental concept. For a recent overview of the literature on the topic and a discussion of how this framework fits with other approaches, see Füssel (2007).

6. This is the framework that was presented in the 2001 Intergovernmental Panel on Climate Change (McCarthy et al. 2001) and further developed since by a number of authors.

7. The index uses principal component analysis to calculate the sensitivity and adaptive capacity subindices, as well as to combine all three indices into the overall vulnerability index. Principal component analysis is a statistical technique that picks the weight given to each component of an index formula in order to best explain the variance in the data. The exposure subindex is from Baettig, Wild, and Imboden (2007) and uses a simple linear formula to combine the underlying variables.

8. The institutional measures are from the Worldwide Governance Indicators Project (Kaufman, Kraay, and Mastruzzi 2008) and include measures of voice and accountability; political stability and absence of violence; and an aggregate governance measure of government effectiveness, regulatory quality, rule of law, and control of corruption.

9. For disaster impact data, see EM-DAT (Emergency Events Database), Centre for Research on the Epidemiology of Disasters (CRED), Université Catholique de Louvain. http://www.emdat.be/Database/terms.html.

10. Thus, while a survey of 928 peer-reviewed articles published in academic journals found no disagreement with the consensus view, public opinion polls show high but far from universal belief in climate change (Oreskes 2004).

11. This section is based on the review of six urban regions' adaptation plans and processes by the Clean Air Partnership (Ligeti, Penney, and Wieditz 2007).

12. There is evidence that even where averages are moving slowly enough and with enough predictability to theoretically provide the necessary time and signals to identify optimal adaptation strategies, the "noise" introduced by extreme weather events and normal climatic variability reduces the value of these signals (Burton and Lim 2005). In other

words, it can be very hard to distinguish a change in the long-term trend from a "bad year."

13. Some argue that individuals are ill-equipped to estimate the risk of low-probability, high-impact catastrophes (Weitzman 2009; Taleb 2007). We have a limited recorded climate history (about 140 years) in which rare events are not well represented. In fact, catastrophe modeling companies accept that the past is an imperfect guide to the future and generate hundreds of thousands of years of synthetic data to model a much broader set of possible catastrophes (Lewis 2007). Statistical approaches, such as extreme value theory, may enable us to underestimate extreme risk.

14. The climate application of this approach, often used in engineering, has been developed in the context of mitigation analysis (Chichilnisky 2000; Azar and Lindgren 2003). See EEA (2007) for a discussion.

15. The flexibility to act in response to signposts (signals that suggest a change in strategy is needed) is limited by the inertia of the system. Thus, it may not be wise to wait for better data on sea-level rise before deciding to halt settlements on a low-lying coastal zone.

16. As pointed out in chapter 2, greater geographic precision of the data does not imply greater accuracy or certainty.

17. Chris West (head of the UKCIP), personal communication.

How ECA's Climate Has Changed and Is Likely to Change Further

Michael I. Westphal

The world is becoming a warmer, wetter place and one where the frequency and magnitude of extreme events is increasing. As the Fourth Assessment Report of the Intergovernmental Panel on Climate Change (IPCC) states: "Warming of the climate system is unequivocal" (IPCC 2007b: 2). Adaptation is unavoidable. This is because past emissions are causing warming that will continue for decades; even if the world stopped producing greenhouse gases (GHGs) today, average temperatures would continue to increase by about 0.6°C over the rest of the century (IPCC 2007b). Continued GHG emissions at or above current rates will induce further changes in the climate system, changes that are likely to be much larger than those experienced in the past century. Thus, while mitigation will not fully substitute for adaptation, the extent of adaptation needed will depend on how much mitigation does in fact occur.

This chapter is based on "Summary of the Climate Science in the Europe and Central Asia Region: Historical Trends and Future Projections" by Michael I. Westphal, a background paper commissioned for this book.

Eastern Europe and Central Asia's Climate Is Already Changing

While interannual (that is, year-to-year) temperatures always vary, a significant increasing trend in temperature can be seen for many sub-regions, particularly the Baltics, the Caucasus, and Central Asia, as well as the northern and eastern parts of the Russian Federation (Westphal 2008). Comparing the mean value for annual temperature in 1901–20 to 1980–2002, one finds that warming has varied from 0.5°C (Southeastern Europe) to 1.6°C (South Siberia) across ECA, and the results are statistically significant for all subregions. There has also been a significant, increasing, year-to-year trend in precipitation over most of Russia, with the exception of the Central and Volga subregion and Baltic Russia, while there have been no significant year-to-year trends in precipitation for the rest of the region.

The change has already shown itself in increased weather-related natural disasters, which have had a large economic impact on the region. Both the number of climate-related natural disasters and the economic losses associated with them have increased, with the clear majority of disasters concentrated in the last two decades (figure 2.1). During this period, drought conditions over much of Eastern Europe

FIGURE 2.1

Natural Disasters in Eastern Europe and Central Asia

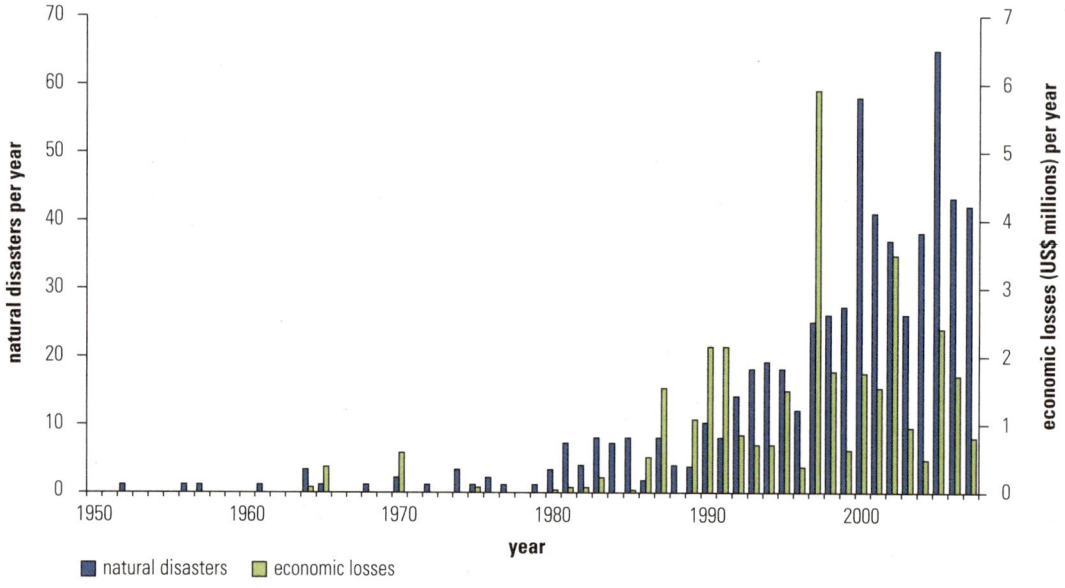

Source: EM-DAT 2008.

Note: Natural disasters include floods, droughts, landslides, extreme temperatures, windstorms, and wildfires. A disaster is defined as an episode leading to 10 or more deaths, affecting 100 or more people, resulting in a declaration of a state of emergency, or leading to a call for international assistance.

and Central Asia have increased markedly, even in regions experiencing increased total annual precipitation.

Many locations within Eastern Europe and Central Asia are in the top three deciles of the global distribution of economic losses per gross domestic product (GDP) for climate-related natural disasters (Dilley et al. 2005). The potential economic loss from natural disasters is particularly severe for the Caucasus and parts of Central Asia (such as Tajikistan, where it is over 70 percent of GDP; Pusch 2004). The two subregions are especially vulnerable to drought because of geographic factors (including high interannual rainfall variability and reliance on snowmelt) and structural factors (heavy economic dependence on agriculture, inadequate hydrometeorological monitoring, and poor water management planning). The 2000–01 drought in the region cost Georgia and Tajikistan an estimated 6 percent and 5 percent, respectively, of GDP (World Bank 2006). While it must be emphasized that a natural disaster is a function of both a natural hazard (climatic event) and an inherent vulnerability, the increase in climate-related natural disasters in Eastern Europe and Central Asia shows that there is an adaptation deficit with respect to current climate variability.

More Change Is Certain—the Question Is Where and How

Adaptation requires an understanding of the potential impacts of climate change on human, economic, and ecological systems. Yet any attempt to estimate such impacts means dealing with a cascade of uncertainties (Schneider and Kuntz-Duriseti 2002). Uncertainty starts with the selection of an underlying emission scenario, which characterizes economic and population growth and energy-use choices. Will the world grow rapidly or slowly? Will developing-country populations soon adopt the consumption habits of high-income countries? What kind of an energy future are we to look forward to?

To assist planning in the face of these uncertainties, the IPCC has developed six socioeconomic scenarios that describe possible trajectories of population and economic growth and the degree of adoption of clean technologies. However, no preferred scenario has emerged, nor has any probability distribution been associated with the scenarios. They are simply different options for the future that imply different carbon emission levels.

There is also uncertainty about the carbon cycle response and the global climate sensitivity, or how the Earth's climate system will respond to the increasing atmospheric carbon dioxide (CO_2) concentration. Climate models work to capture the highly complex interaction of many different influences (for example, ocean, atmosphere, and cryosphere) on the weather. As a result, they differ in their projections. The IPCC uses a set (ensemble) of global climate models (also called Global Circulation Models, or GCMs) instead of a single one. The variation of results across models gives an estimate of uncertainty.

Local changes in climate are also affected by local features, such as mountains, which are not well represented in global models because of their coarse resolution. Capturing local characteristics requires downscaled models (box 2.1). Regional climate models (RCMs) provide projections at a much finer scale (typically using cells measuring 50 km by 50 km as opposed to 300 km to 500 km for global models) for limited areas, taking their input from GCMs at their boundaries. However, greater precision does not guarantee greater reliability.

Finally, uncertainty is magnified with attempts to estimate impacts (such as those on ecosystems, health, agriculture, housing, and the economy), since this requires developing another set of models that include sectoral information and socioeconomic behavior and making assumptions as to people and systems' capacity for adaptation (EEA 2007).

But uncertainty is no excuse for inaction. As indicated in chapter 1, countries must develop adaptive capacity rather than seek to adapt to one particular outcome. As the next section shows, there is no doubt that change is coming. While it is a good idea to improve and refine projections, it would be a very poor strategy to do nothing until projections become "more precise," however that is defined.

Climate Projections: How Is the Eastern Europe and Central Asia Region Likely to Be Affected?

There is consensus about broad climate trends over the twenty-first century, particularly if we limit ourselves to general qualitative assessments (for example, milder winters and hotter summers). This book, however, goes a step further: the IPCC ensemble of models was used to generate projections for the period 2030–49 (the most relevant one for adaptation policy), assuming a world of rapid economic growth, slow population growth, and very high, but cleaner, energy use (otherwise known as the A1B scenario).[1]

Given Eastern Europe and Central Asia's tremendous climate diversity (from polar to Mediterranean), the region was divided into six subregions plus Russia, itself divided into another seven, for a total of 13 subregions based on a combination of political boundaries and agro-ecological zones (map 2.1). A regional overview is pre-

BOX 2.1

General Circulation Models and Climate Downscaling

Most of the data presented in this report are from General Circulation Models (GCMs). GCMs are spatially explicit, dynamic models that simulate the three-dimensional climate system using as first principles the laws of thermodynamics, momentum, and conservation of energy, and the ideal gas law. GCMs divide the world into a grid, and each equation is solved at each grid cell across the entire globe, at a fixed time interval (usually 10 to 30 minutes), and for several layers of the atmosphere. Due to the computational burden, GCMs typically have spatial resolutions of 1 degree to 4 degrees ($\sim$100–400 km^2). The coarseness of the spatial resolution means that aspects of those climate dynamics that have smaller spatial scales—such as topography, clouds, and storms—are imperfectly incorporated and averaged over the entire grid cell (Wilby et al. 2009). Generally, climate models perform better in projecting temperature than precipitation and mean changes rather than extreme events (Solomon et al. 2007).

Climate downscaling in regard to climate change projections is an umbrella term that includes two approaches for enhancing precision but not necessarily improving accuracy. *Dynamic downscaling* generates Regional Climate Models (RCMs), while *empirical downscaling* relies on locally observed statistical relationships (Wilby et al. 2009). Because both rely on data and boundary conditions from GCMs, it is pointless to downscale where there is limited confidence in the GCMs (Schiermeier 2004). Downscaling should be undertaken only in regions where the GCMs are in general agreement, which signals greater reliability.

RCMs simulate climate dynamically at very fine scales (10 km to 50 km). The atmospheric fields simulated by a GCM (surface pressure, temperature, winds, and water vapor) are entered as boundary conditions for the RCM, and the "nested" RCM then simulates the smaller-scale climate. RCMs have been shown to realistically simulate regional climate features, such as precipitation, extreme climate events, and regional-scale climate anomalies, such as those associated with the El Niño Southern Oscillation (Wang et al. 2004). However, RCMs are sensitive to the errors of the "mother" GCM models, which specify the boundary conditions and the choice of initial conditions, such as soil moisture.

Empirical downscaling relies on determining statistical relationships between large-scale atmospheric variables (for example, strength of airflow and humidity) with local response variables, such as daily precipitation. Changes in those large-scale variables under climate change as simulated by GCMs can be translated into changes in local predictor variables and, thus, outcomes.

A plethora of downscaling software is available; however, access to data on predictor variables for calibration presents a major impediment to their use. Empirical downscaling relies on having good observational data and accurate predictions of the relationship of the local variables to large-scale forcing, as well as knowledge of how that relationship may be altered by climate change. In one global study of daily precipitation, empirical downscaling performed relatively poorly in near-equatorial and tropical locations, but it adequately reproduced seasonal precipitation and the phase of daily precipitation in mid-latitude locations (Cavazos and Hewitson 2005).

Source: Westphal 2008.

MAP 2.1

Eastern Europe and Central Asia Subregions

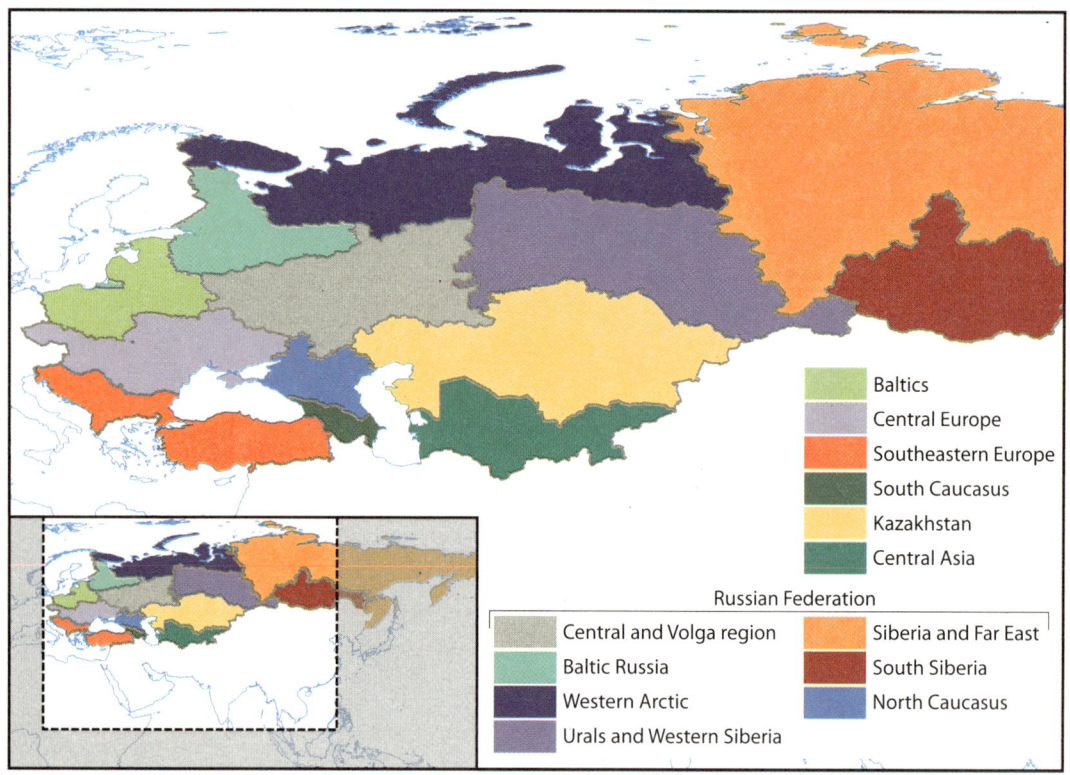

Source: Westphal 2008.

sented here, but climate summary sheets are available for each of the 13 subregions in the background paper (Westphal 2008).

Warmer Everywhere: Fewer Frost Days, More Heat Waves

The ensemble of GCMs projects continued warming everywhere, with fewer frost days and more heat waves. There is complete model concordance on the direction of these changes (box 2.2). Map 2.2 summarizes the projections: the projected increase in mean annual temperature in Eastern Europe and Central Asia ranges from 1.6°C to 2.6°C by the middle of this century, with a gradient of increasing temperature change with more northern latitudes. The northern parts of the region will have greater temperature changes in the winter, while the more southerly parts will show greater warming in the summer months. The number of frost days is projected to decrease by 14 to 30 days, with the greatest decrease occurring in the Baltic subregion. Furthermore, the number of hot days is projected to increase by 22 to 37 days per year by 2030–49, and the greatest increases in heat wave

BOX 2.2

The Skill of Models in Simulating Present Climate in Eastern Europe and Central Asia

The credibility, or reliability, of climate models can be tested by comparing model-generated climate simulations for the current period against currently observed climate. Such an exercise reveals that regional climate models perform better generally than global climate models, but that performance varies across Eastern European and Central Asian subregions, with Central Asia most poorly served.

Europe

Existing GCMs exhibit either positive or negative biases for temperature in the summer, while most have a cold bias in the winter, particularly for Northern Europe (meaning that models tend to project colder than actual temperatures in the winter). Biases in temperature vary considerably in Europe, both spatially and temporally. Precipitation biases for Europe are smaller than those for most regions of the world. Most GCMs overestimate rainfall from autumn to spring in Northern Europe, but many also overestimate summer rainfall. In the Southern Europe and Mediterranean region, the median simulated annual precipitation is very close to observation, although models differ in the sign of the small bias.

RCMs for Europe capture the geographical variation of temperature and precipitation better than global models but tend to simulate conditions that are too dry and warm in Southeastern Europe in summer. Most but not all RCMs also overestimate the interannual variability of summer temperatures in Southern and Central Europe.

Asia

For Russia west of the Urals, most models have negative temperature and positive precipitation biases: they underestimate temperatures but overestimate precipitation. GCMs typically perform poorly over Central Asia due to the topography. Models, even RCMs, tend to overestimate precipitation over arid and semi-arid areas in the northern part of Central Asia. The precipitation biases range from –58 percent to +24 percent over the ensemble of models over all seasons. RCMs for Central Asia are much less developed than those for Europe.

Source: Westphal 2008.

duration are expected in the North Caucasus.[2] In general, daily minimum temperatures are projected to increase faster than daily maximum temperatures, narrowing intra-daily temperature ranges (Parry et al. 2007).

Our projections are consistent with the results of downscaled models available for the region. In Europe, a team of 21 European

MAP 2.2

Projected Changes in Annual and Seasonal Temperature by Mid-Century

a. Change in mean annual temperature

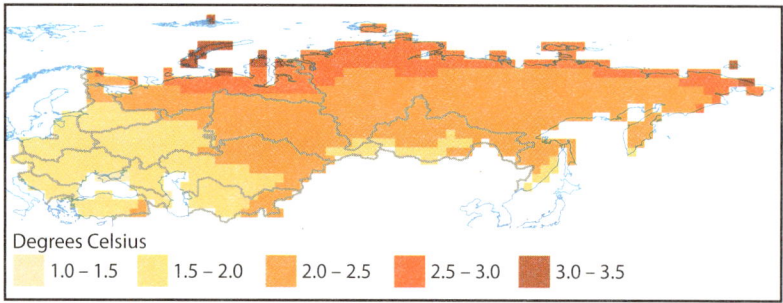

b. Change in winter (DJF) temperature

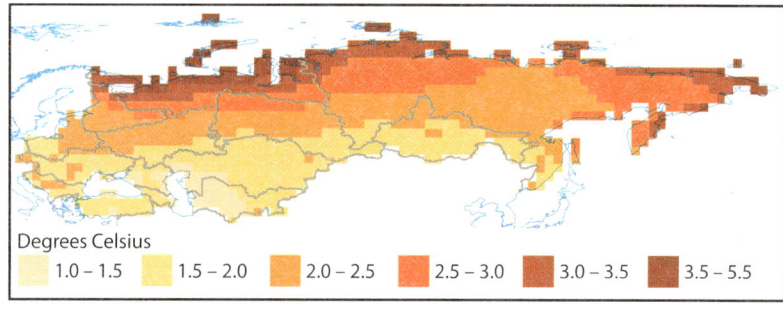

c. Change in summer (JJA) temperature

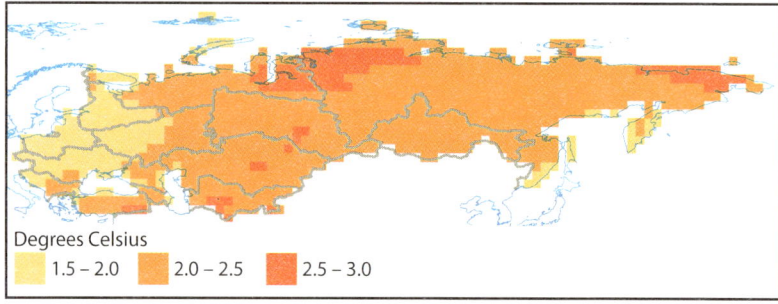

research groups undertook an interdisciplinary project—PRU-DENCE (Prediction of Regional scenarios and Uncertainties for Defining EuropeaN Climate change risks and Effects)—to provide high resolution climate change scenarios for Europe at the end of the twenty-first century using RCMs (Christensen et al. 2007). In terms of extreme temperature events, the PRUDENCE RCM projections reiterate the patterns seen in the GCM projections presented here. By the end of the twenty-first century, Central Europe (roughly corresponding to the same subregion of this study) is projected to experience the same number of hot days (>30°C) as Spain and Sicily.

MAP 2.2
Continued

d. Change in number of frost days

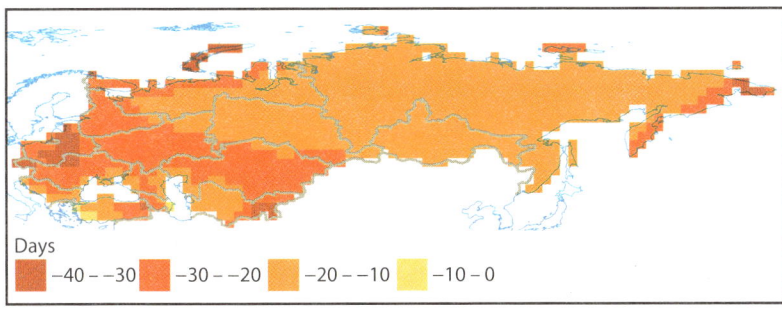

Days

−40 – −30	−30 – −20	−20 – −10	−10 – 0

e. Change in heat-wave duration index

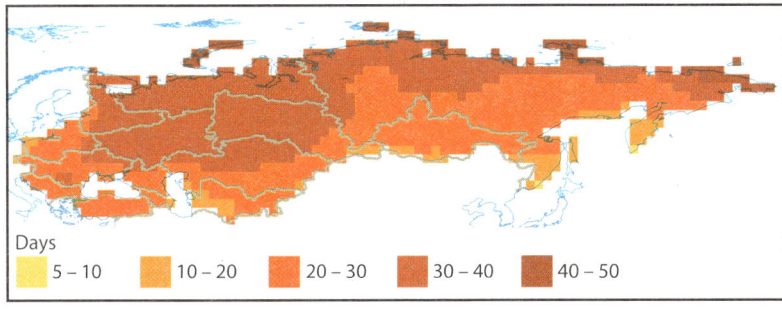

Days

5 – 10	10 – 20	20 – 30	30 – 40	40 – 50

Source: Westphal 2008.
Note: DJF = December, January, February. JJA = June, July, August. All projections are based on the IPCC A1B climate scenario, and combine the results of eight general circulation models. The projections indicate the difference between average temperatures in the period 2030–49, compared to the period 1980–99. See note 2 for definition of heat wave in (e).

An RCM has also been developed for parts of Russia and Central Asia (Shkolnik, Meleshko, and Kattsov 2007). In winter, the model projects a decrease in temperature variability and cold extremes, while in the summer, extremely high daily temperatures are projected to increase at a faster rate than the rest of the temperature distribution. As for heat wave duration, the model projects the most severe increases for Central Asia, Kazakhstan, the Urals, and Western Siberia.[3]

The following section summarizes expected general trends for the region. Subregional trends are summarized in annex table 2.1, while more disaggregated and detailed projections are available in Westphal (2008).

A Wetter, Rainier North and East, and a Drier South

The GCMs project a wetter north and east and a drier south (map 2.3). By mid-century, mean annual precipitation will increase in most of the Russian subregions (5 percent to 11 percent), with the North Caucasus

MAP 2.3

MAP 2.3

Projected Changes in Annual and Seasonal Rainfall by Mid-Century

a. Change in mean annual rainfall

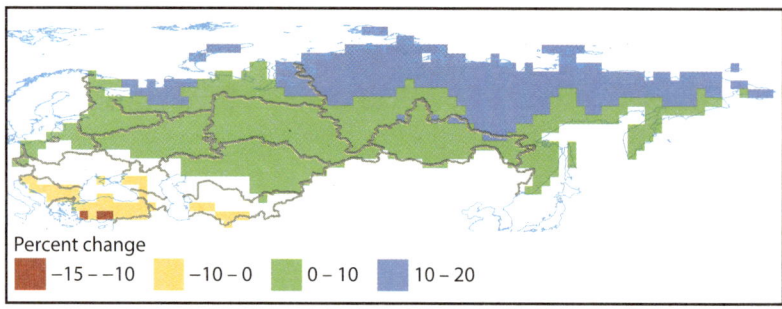

b. Change in winter (DJF) rainfall

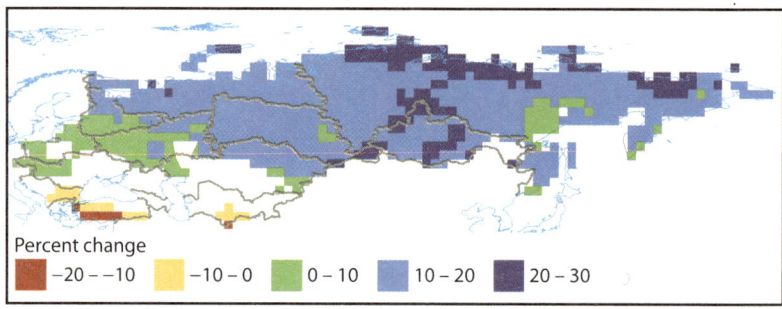

c. Change in spring (MAM) rainfall

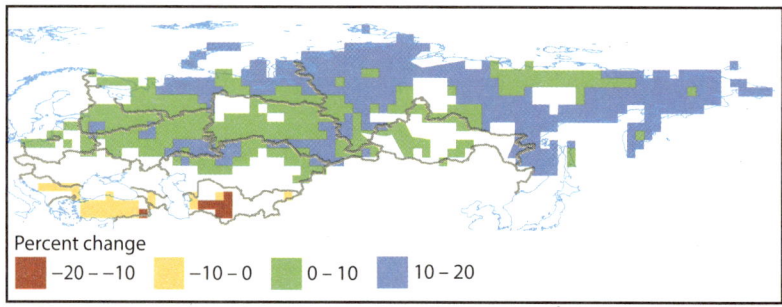

the only anomaly (–2 percent). The increase is most pronounced in Siberia and the Far East. In addition, winter precipitation in Russia is projected to increase more substantially than precipitation during the other seasons (9 percent to 18 percent, excluding the North Caucasus).

For all of the Russian subregions, there is strong model agreement in mean annual and winter precipitation; the situation for summer precipitation is inconsistent, with the exception of South Siberia. However, by the end of the century, there are clear consistent trends for increased precipitation in most of Russia for the summer months (Parry et al. 2007; Kattsov et al. 2008). For the rest of Eastern Europe and Central Asia, the most consistent trends are: an increase in win-

MAP 2.3
Continued

d. Change in summer (JJA) rainfall

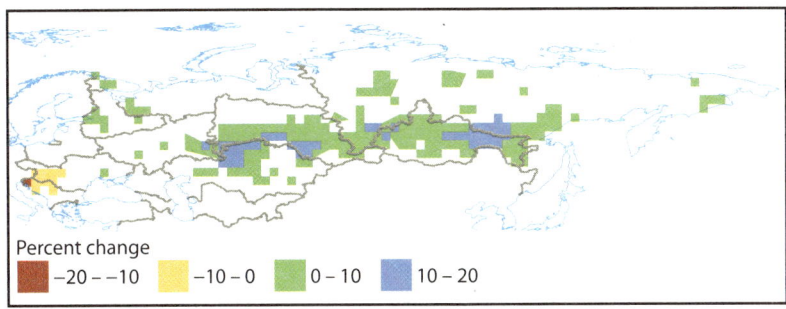

Percent change

▮ −20 – −10	▮ −10 – 0	▮ 0 – 10	▮ 10 – 20

e. Change in autumn (SON) rainfall

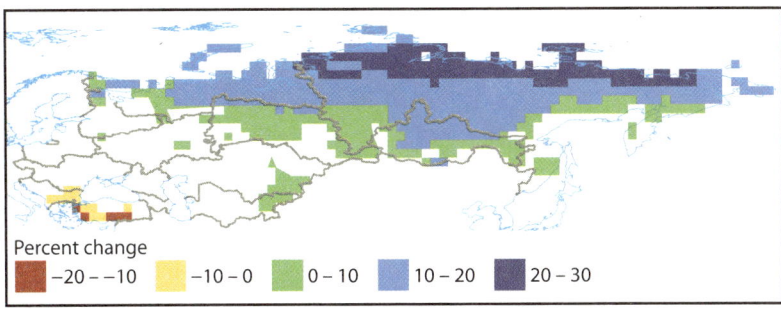

Percent change

▮ −20 – −10	▮ −10 – 0	▮ 0 – 10	▮ 10 – 20	▮ 20 – 30

Source: Westphal 2008.
Note: DJF = December, January, February. JJA = June, July, August. MAM = March, April, May. SON = September, October, November. All projections are based on the IPCC A1B climate scenario, and combine the results of twenty general circulation models. The projections indicate the difference between average precipitation in the period 2030–49, compared to the period 1980–99. Blank areas indicate where fewer than two-thirds of the models agree on the sign of the change.

ter (9 percent) and spring (5 percent) precipitation in Kazakhstan by mid-century and a decrease in precipitation in Southeastern Europe (–6 percent for the annual mean). The near-term picture of summer precipitation in Southeastern Europe is inconsistent across models, although end-of-century projections show consistent trends of decreasing precipitation (Parry et al. 2007). There is strong model disagreement for annual and seasonal precipitation on average for the Baltics, the Caucasus, Central Asia, and Central Europe.

Finally, the models project that the interval between rainfall events will decrease in the north and east and increase in the south and west, with the greatest magnitude in Southeastern Europe (maximum consecutive dry days [CDD] increasing by 5 days) and Asian Russia (CDD decreases by 4 days in South Siberia). Runoff, a measure of water availability, is projected to decrease everywhere but Russia (map 2.4). The most dramatic decrease will likely occur in Southeastern Europe (–25 percent) and result in increased drought conditions (Milly, Dunne, and Vecchia 2005; Milly et al. 2008). The

MAP 2.4

Projected Changes in Consecutive Dry Days, Runoff, and Rainfall Intensity by Mid-Century

a. Change in consecutive dry days

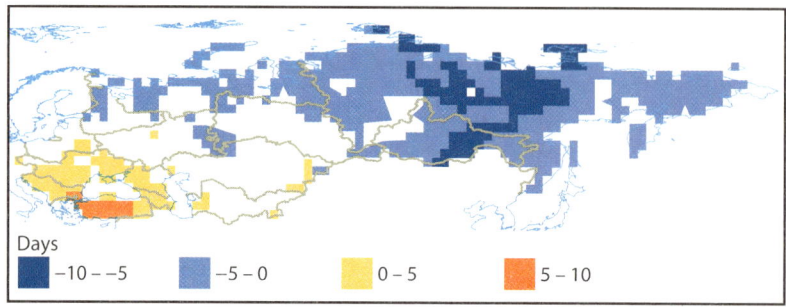

b. Change in runoff (a measure of water availability)

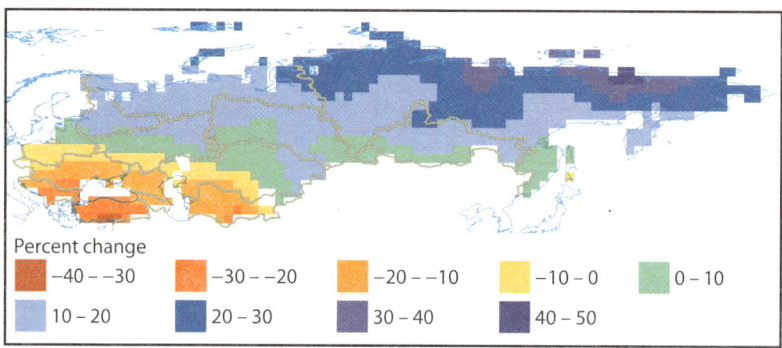

c. Change in daily rainfall intensity

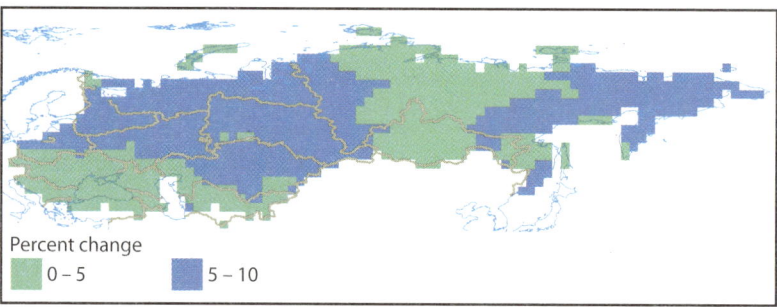

d. Change in maximum total rainfall during a 5-day period

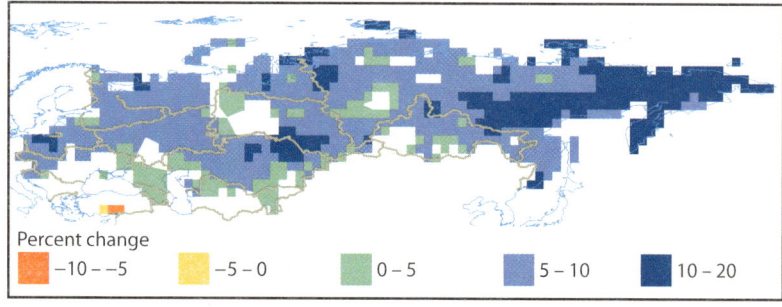

Source: Westphal 2008.
Note: All projections are based on the IPCC A1B climate scenario, and combine the results of eight general circulation models. The projections in (a) and (b) indicate changes in the period 2041–60, compared to the period 1900–70. The projections in (c) and (d) indicate changes in the period 2030–49, compared to the period 1980–99. Rainfall intensity, in (c), is measured as the ratio of annual total rainfall to the number of days during the year when rainfall occurs. Blank areas indicate where fewer than two-thirds of the models agree on the sign of the change.

net impact is less clear in Russia: precipitation and runoff are projected to increase, but so are temperatures and heat waves, speeding evaporation and reducing water availability. Most of the precipitation increase in Russia is expected in the winter, and while low runoff is often used as a proxy for drought, the runoff indicator is an annual average that masks temporal variation.

It is still possible for higher summer temperatures to offset precipitation increases and lead to periodic drought conditions in the future. One projection of the Palmer Drought Severity Index shows an increase in drought conditions over the course of the twenty-first century over much of Russia, with the exception of far northeastern Siberia (Dai, Trenberth, and Qian 2004; Aiguo Dai, personal communication). Russia will likely receive more precipitation, but whether this excess can be captured and put to use is uncertain.

When It Rains, It Pours—Everywhere

Throughout the entire region, the models are unequivocal: precipitation intensity will increase, ranging from 2 percent to 6 percent (map 2.4). While this may not seem significant, these are mean values and depend on local hydrology and topography. This increase in precipitation intensity could have significant repercussions for water storage systems, sanitation, and flood management. With the exception of Southeastern Europe, most models project an increase in precipitation from extreme storm events (2 percent to 9 percent) and an increase in the maximum amount of precipitation over a 5-day period. The PRUDENCE RCMs project heavy winter precipitation increases in Northern and Central Europe and decreases in the south. In the summer, the zone of heavy precipitation shifts to Northeast Europe. Extreme wind speeds (winter storms) are projected to increase over Central Europe (Beniston et al. 2007).

The projections for extreme precipitation cannot be translated directly into flood projections; detailed local-scale impact models, incorporating topography and specifics of hydrology, are needed. However, if a region is currently experiencing significant flooding and if no adaptation measures (flood mitigation) are enacted, then one can assume that an increase in extreme precipitation will result in more flooding. Whether this results in more disasters depends on whether vulnerability (for example, land-use planning, the population in the floodplain, the existence of early warning systems, and institutional capacity) remains constant.

ANNEX TABLE 2.1

General Climate Trends in the Eastern Europe and Central Asia Subregions

Subregion	Current trends and weather-related events	Projected temperature rise by 2050	Mean annual precipitation	Runoff	Rainfall intensity and variability	Interval between wet days	Heat waves
Baltics	Warming trend over the past century; flood damage significant	1.6°C, warmer winters, decrease in frost days	Unclear	South: decrease; north: increase	Increased intensity	Unclear	Increase
Central Asia	Warming trend over the past century; droughts and landslides in some parts	2.0°C, decrease in frost days	Unclear	Decrease	Increased intensity	Unclear	Increase
South Caucasus	Warming trend accelerating in past 20 years; droughts and landslides in parts	1.7°C, warmer summers, decrease in frost days	Unclear	Decrease	Increased intensity and more variability	Increase	Increase
Central Europe	Warming in the past 20 years, but no trends in precipitation	1.7°C, decrease in frost days	Unclear	Decrease (median 13%)	Increased intensity and more variability	Increase	Increase
Kazakhstan	Warming over past century	2.0°C, decrease in frost days	Increase (4–9%)	Slight increase	Increased intensity	Unclear	Increase
Southeastern Europe	No trends, but vulnerable to floods and drought	1.8–2.1°C, decrease in frost days	Decrease (−6%) except unclear in summer	Decrease (25%)	Increased intensity	Increase	Increase
Russian Federation regions							
Baltic Russia	Significant flood and landslide damage in some parts	1.9°C, decrease in frost days	Increase (6%), more in winter and spring	Increase (13%)	Increased intensity	Unclear	Increase
Central and Volga	No trends; flooding significant	1.9°C, warmer winters, decrease in frost days	More in winter and spring	Increase (7%)	Increased intensity	Unclear	Increase
North Caucasus	Increasingly wet over the past century	1.6°C, decrease in frost days	Unclear	Decrease (12%)	Increased intensity and more variability	Decrease	Increase
Siberia and Far East	Significant warming and wetting in the past century	2.4°C, decrease in frost days	Increase (11%), particularly in winter (17%)	Increase (22%)	Increased intensity	Decrease	Increase
South Siberia	Warming and wetting trend over the past century; floods and landslides	2.1°C, decrease in frost days	Increase (8%)	Increase	Increased intensity	Decrease	Unclear
Urals and West Siberia	Significant wetting in past century; floods and landslides	2.2°C, decrease in frost days	Increase (9%), particularly in winter (15%)	Increase (10%)	Increased intensity	Unclear	Increase
Western Arctic	Significant warming and wetting in the past century	2.6°C, even more in winter, decrease in frost days	Increase (10%), particularly in winter (16%)	Increase (17%)	Increased intensity	Decrease	Increase

Source: Derived from Westphal 2008.

Notes

1. Of the four "pillar" emission scenarios (A1, A2, A1B, and B2) (Solomon et al. 2007), only projections for the mid-range A1B scenario are shown. There are no significant differences across scenarios in their warming projections until 2030; moreover, even by mid-century, the variation among climate models for a given emission scenario tends to be greater than the variation among multi-model means calculated for each emission scenario. For a full discussion of the methodology and model used, see Westphal 2008.
2. Heat wave duration is a relative measure of the number of consecutive days that have a daily maximum temperature at least 5°C greater than the historical normal daily maximum temperature.
3. In this particular model, heat wave duration is defined as the sequence of days with daily maximum temperatures exceeding the local 90th percentile of previous summers' maximum temperature distributions.

CHAPTER 3

Human Health: The Most Basic Vulnerability

Tamer Rabie, Safinaz el Tahir,
Tereen Alireza, Gerardo Sanchez,
Katharina Ferl, and Nicola Cenacchi

Countries of all income levels are vulnerable to natural forces, as was amply demonstrated by Hurricane Katrina in the United States in 2005 and by the heat wave in Europe in 2003. When extreme weather destabilizes the balance between natural and human systems, protective structures and institutions can quickly break down—particularly those that are already weak or stressed—eventually threatening human lives and well-being.

In Eastern Europe and Central Asia (ECA), the most urgent health issues arising from climate change relate to already vulnerable populations: persons who are elderly, ill, very young, displaced, or marginalized. When extreme weather combines with political instability and civil strife, the numbers of people facing serious health emergencies can multiply, as experienced in post-independence Georgia in the 1990s. Individuals who live in substandard housing, as do many Roma, will be hardest hit by floods and heat waves.

This chapter is based on a background paper prepared for this book, "The Health Dimension of Climate Change," by Tamer Rabie, Safinaz el Tahir, Tereen Alireza, Gerardo Sanchez, Katharina Ferl, and Nicola Cenacchi, and was drafted by Tim Carrington.

Long-term threats to health under a changed climate may be less easy to measure or attribute than those resulting from extreme weather events such as floods or droughts—but they are also important. A more stressed agriculture sector will translate into higher rates of malnutrition and increased susceptibility to disease. Families that depend on rain-fed agriculture will be affected by shifts in precipitation and may migrate to seek improved livelihoods, thereby increasing the numbers of people underserved by local health systems. Water degradation from a variety of sources will expose more people to dengue fever and diarrheal diseases.

What follows is an examination of two categories of health risk:

- the immediate and direct threats occasioned by warmer, wetter weather, with more climatic extremes, and

- the setbacks brought on by the consequences of and adjustments to climate changes, including interrupted livelihoods, migration and temporary displacement, and inadequate nutrition.

Warmer and More Extreme Weather Brings New Threats and Exacerbates Others

Extreme weather events, such as floods and droughts, pose the most immediate and obvious health risks—and projections indicate these events will become more intense and frequent. In addition, the threats arising from extreme events are sometimes aggravated by parallel crises, including civil strife, breakdown in health systems, and institutional collapse.

Floods

Floods, which account for half the world's natural-disaster fatalities, constitute a multi-pronged assault on the human system. From 2000 to 2007, ECA's 10 most severe floods—in the Russian Federation, Turkey, Romania, Poland, and Tajikistan—accounted for nearly 500 casualties (table 3.1). Deaths from drowning or collapsing structures were compounded by the landslides that frequently accompany floods. Evacuations, particularly those involving hospital patients and other vulnerable groups, are enormously stressful and increase the risk of heart attack.

But some of the health impacts are less immediate and less obvious. Post-traumatic stress, increased poverty, compromised nutrition,

TABLE 3.1

A Rising Tide of Flooding Episodes

Year	Country	Location	Casualties
2002	Russian Federation	Novorossiysk	167
2002	Russian Federation	Stavropol, Krasnodar, Karachaevo-Cherkessia, Ingushetia, Adygea, Chechnya, Kabardino-Balkaria, North Ossetia-Alania, Dagestan	91
2006	Turkey	Çinar, Bismil	47
2002	Turkey	Rize, Çorum, Yozgat, Kars and Muş provinces	34
2005	Romania	Harghita, Mureş, Dolj, Bacău, Vrancea, Galaţi, Brăila, Bistriţa, Gorj, Suceava	33
2006	Romania	Arbore, Bistriţa, Maramureş, Arad	30
2001	Poland	Małopolskie, Świętokrzyskie, Dolnośląskie, Oploskie, Śląskie, Warmińsko-Mazurskie, Podlaskie, Gdańsk, Słupsk regions	27
2005	Romania	Alba, Tulcea, Giurgiu, Vrancea, Bacău, Brăila, Galaţi, Vrancea, Ialomiţa	24
2002	Tajikistan	Dasht, Langar	24
2007	Tajikistan	Asht district	21

Source: EM-DAT 2008.
Note: Table includes the 10 most lethal, in number of people killed, flood events in ECA recorded in EM-DAT for 2000–07.

and interrupted livelihoods all affect human health without appearing immediately as illnesses or injuries caused by extreme weather. Long-term displacement of people and permanent migration from flood-damaged residences go hand-in-hand with lower living standards and increased vulnerability.

Georgia's experiences over the past 30 years demonstrate the ways that extreme weather combines with institutional weakness and civil strife to further lower the quality of life for thousands of citizens. Between 1987 and 1989, earthquakes, floods, and landslides caused the displacement of 20,000 people in the Svanetia and Ajara regions (UNHCR 2004). At the same time, because of the disruptions during the unraveling of the former Soviet Union, Georgia suffered a severe shortage of medical supplies and required international assistance (OFDA 1992). Civil strife added refugees fleeing violence to those uprooted by natural disaster, leading to a number of crowded, unhealthy, and highly insecure temporary settlements. In 1993, thousands of people were driven out of makeshift homes, and again in 1998, by which time the population of internally displaced reached 40,000 (OFDA 1999). Extreme weather events—in a context of political instability, institutional weakness, and poverty—became a major contributor to increased poverty, insecurity, and vulnerability.

While Georgia emerges as the most vulnerable for the period 1980–2000, measured by the mortality rate among those exposed to floods, it was not alone. The Czech Republic was the second most vulnerable for the same period, followed by the Slovak Republic and Moldova (UNDP DRI).

Following the immediate damage and trauma of a flood—but well before the long-term effects of displacement and loss of income run their course—there may be a wave of health risks stemming from waterborne illnesses as sewage, industrial wastes, and agricultural runoff flow into human settlements and degrade the water supply (box 3.1).

The upheaval caused by a flood and the resulting loss of homes, possessions, and livelihoods leaves people strained and exhausted, often suffering from post-traumatic stress disorder and depression. The fallout can be serious following floods that displace and destroy on a wide scale. For example, when Poland's Oder River flooded in 1997, it affected 86 cities and towns, 875 villages, and 450,000 farms (OFDA 1998), with an estimated overall economic cost of US$3.5 billion (EM-DAT 2008). The Federation of Red Cross and Red Crescent Societies reported 50 suicides linked to the disaster in a two-month period (Hajat et al. 2003). High levels of physical and emotional stress affect a host of bodily systems, complicating pregnancies and raising the risk of heart disease.

Heat Waves

Heat waves have an immediate impact on public health, often aggravating a variety of health conditions and bringing about unhealthy changes in water or air quality. Researchers have found that during an extended period of intense heat, the number of deaths rises above established seasonal norms. These are considered "excess deaths"—those specifically attributed to the effects of intense heat.

Cities intensify heat waves because traffic, buildings, and sparse vegetation all increase temperatures further. In 2001 in Moscow, 276 deaths in excess of the multi-year average were attributed to a nine-day heat wave. That same summer, heat waves may have caused hundreds of deaths in Croatia, the Czech Republic, and Slovenia. The latest estimates for the pan-European heat wave of 2003 point to 70,000 deaths (Robine et al. 2008). The 2003 heat wave was the most dramatic in recent history, but a number of fatal heat waves

BOX 3.1

With Every Flood, a Risk of Disease

Flooding, apart from causing drowning and injury from collapsing structures, introduces a host of illnesses as water supplies are contaminated with sewage and wastewater from farms and factories. Poorly maintained water systems and inherited environmental degradation add to the risks.

The following flood-related illnesses, already present in the ECA region, are projected to become more frequent threats:

- *Dysentery,* an infectious disease caused by the bacterium *Shigella dysenteriae,* is a common threat in floods. In Tajikistan in 1992, flooding combined with displacement from civil unrest put hundreds of people at risk, resulting in higher childhood mortality in two villages.

- *Typhoid fever,* an infectious disease carried by feces and urine, is caused by the bacterium *Salmonella typhi.* In May 1996, following heavy rains and flooding in Tajikistan, a poorly maintained sewage system came under additional stress and contaminated the water supply. In the ensuing typhoid fever outbreak, 7,516 cases were reported in a month's time, one-third of them in children under age 14. As in Georgia, simultaneous stresses on institutions and infrastructure from the flooding and prior weaknesses combined to worsen the health crisis. About 50 health clinics and schools were damaged by the floods. The toilet system of a major hospital was inundated, further spreading the dangerous bacterial contaminant. Prior conditions added to the population's vulnerability once the flood came. Amid civil violence, public funding of health facilities tapered off, leaving the system short of diagnostic supplies and drugs for treatment of infectious diseases. In 1995, soap had become largely unavailable and chlorination of the water supply had been halted due to a lack of materials. In some parts of the country, people had begun using open canals for their water supply, but hundreds of these were ruined in the floods.

- *West Nile virus,* which is highly dangerous for the elderly, is spread primarily by mosquitoes, whose larvae thrive in the pools of standing water left by flooding. An outbreak of the disease followed 1999 floods in the Czech Republic, when *Aedes* mosquitoes proliferated in affected areas. Europe's largest recorded outbreak occurred in Bucharest, Romania, in 1996 and showed that urban areas also were vulnerable, with larvae multiplying in the flooded basements of buildings.

- *Tahyna,* a virus that breeds in flooded areas, was detected in the Czech Republic following three separate episodes of flooding.

- *Leptospirosis,* a once rare infectious disease carried by rodents and other animals, spreads through contact with moist soil, mud, vegetation, or contaminated water. The Czech Republic, Russia, and Ukraine have experienced outbreaks following floods.

Other waterborne diseases, including cholera, hepatitis A, and salmonella, have also surfaced in the region following flooding episodes.

Source: Rabie et al. 2008.

TABLE 3.2

Heat Waves Add to Illnesses

Year	Heat wave temperature record (°C)	Country (location)	Number of heat wave–related morbidities[a]
2005	36	Romania (Bucharest)	500
2000[b]	46	Turkey	300
2000	35	Croatia (Zagreb, Split, Osijek, Rijeka)	200
2006	36	Romania	200
1996	40	Romania	200
2000	43	Romania (Bucharest, Bechet)	100
2007	40.3	Slovak Republic	89
2000	42	Serbia and Montenegro	70
2007[c]	45.5	Bulgaria	50

Source: EM-DAT 2008.
a. Figures reported in EM-DAT as number of people injured and people suffering from physical injuries, trauma, or an illness requiring medical treatment as a direct result of a disaster.
b. Heat wave associated with drought event.
c. Heat wave associated with wildfires and drought event.

have occurred in Central and Southeastern Europe over the last 10 years (table 3.2).

The following categories show the ways that periods of intense heat generate new health threats or undermine the body's capacity to manage existing conditions:

- *Heat stroke* is a severe condition in which, under excessive exertion, the body ceases sweating. This causes body temperature to rise to dangerous levels and can result in fainting, organ failure, and death.

- *Heat cramps and heat exhaustion* occur when the body sweats so much that the concentration of salt in the body becomes dangerously low. The condition can increase the heart rate and lead to heat stroke if left untreated. Infants and small children are at risk because their fluid reserves are smaller than those of adults. Elderly persons, who may eat and drink little because of weak appetite or take medications that leave them more prone to dehydration, are also at risk.

Exacerbation of existing conditions is a major risk, since many cardiovascular, cerebrovascular, renal, respiratory, and psychological conditions are sensitive to heat. For example, during 17 heat waves in the Czech Republic over an 18-year period, there was a 13.6 percent increase in cardiovascular mortality (Kysel and Huth 2004).

Studies in Croatia and Uzbekistan have found weakened performance of heart patients during times of extreme heat.

Stressed infrastructure can compromise utility service delivery and thereby worsen health conditions; as chapter 6 explains, ECA's inherited stock of Soviet-era infrastructure and poorly ventilated housing is vulnerable to atypical heat. Heavy use of air conditioning alleviates risks for people who can access or afford it, but it strains power supplies and may lead to outages. Electricity outages also limit water access for many people, potentially setting off a cascade of other impacts.

Extreme heat can also lead people to engage in riskier behaviors, such as swimming in open canals, rivers, or lakes, leading to deaths that would not occur in less extreme summer weather. Pollution, smog, and fires—which often intensify during a heat wave—lead to greater-than-normal cardiovascular problems and deaths. Because of extreme heat and a lack of rainfall in 2007, wildfires proliferated in Southeastern Europe, causing dozens of hospital visits as well as a number of deaths.

Droughts

Droughts, depending on their severity and duration, present a variety of health risks. A severe drought in 2000 and 2001 in Tajikistan and Uzbekistan cut the availability of drinking and irrigation water and led to slow, chronic forms of malnutrition as households eliminated meat and dairy products from their diet (WHO Europe 2001).

The drought that hit Moldova in the summer of 2007 offers a well-documented case of health impacts. A survey by the World Food Program and the Food and Agriculture Organization of the United Nations estimated that the crisis affected 84 percent of the country's arable land, leading to estimated economic losses of US$407 million from crop failures and livestock deaths (quoted in UN Moldova 2007a).

Strains on ordinary citizens were evident. A household survey showed that 72 percent of the households interviewed were worried about having enough food (UN Moldova 2007b). Of households with three or more children, 59 percent reported that they ate differently, with some foods they formerly counted on now unavailable. Nearly 40 percent of the households surveyed said their water source was dried up or at least damaged by the drought conditions.

Changing Averages: Malaria, Allergies, and Algal Blooms

Diseases associated with warmer weather will probably become more prevalent in ECA, and some have already surfaced. A major concern is malaria. Largely eradicated from Europe, malaria has returned to the Caucasus and Central Asia, with weather-related events raising disease levels. For example, mudslides in 1997 elevated the prevalence of malaria in Azerbaijan, increasing the number of breeding sites considerably (WHO Europe 2005). Malaria is also endemic in Turkey and Tajikistan, where the Roll Back Malaria program was introduced in 1998.

Warmer average temperatures will also increase pollen-related allergies, particularly in Central Europe where the ragweed *Ambrosia* is more highly concentrated than in most other regions of the world. Pollen concentration increases with higher temperatures and higher ambient concentrations of carbon dioxide.

Changing averages are affecting health in other ways. Warmer, wetter weather is changing conditions in the Baltic Sea, with ramifications for human health. One process under way is eutrophication, involving an increase in nutrients (usually nitrogen and phosphorous) in the sea. The process triggers algal blooms that lead to a degradation of environmental quality (HELCOM 2007). One category of algal blooms, cyanobacteria (present in the Baltic for decades), has recently increased in duration, frequency, and aggregate biomass (Bianchi et al. 2000). Resulting toxins trigger gastrointestinal illnesses and liver damage in cases of persistent exposure, and ingesting contaminated water has killed cattle and pets (WHO 2003). The toxins are a risk to human health as well and have been linked to carcinomas in China (Chorus and Bartram 1999). Climate change models project two types of impacts in the Baltic Sea—increased freshwater runoff into the Baltic from increased precipitation and flooding, plus warmer sea temperatures—both of which can contribute to increased cyanobacteria (HELCOM 2007).

The Climate Change–Health Outcome Matrix

Table 3.3 distills the findings and projections of an extensive literature on the relationship of current climate and climate change with human health.

Vulnerability from Climate-Driven Migration: The Health Perspective

In rural areas, livelihoods depend directly on climate-sensitive resources, particularly water, and settlements are highly exposed to weather extremes. Households that earn income from farming and livestock activities have historically resorted to seasonal or indefinite migration when conditions become too harsh or too precarious. Families that are

TABLE 3.3

Health Consequences of a Changing Climate: Direct and Indirect

Exposure / Outcome	Direct impacts — Extreme weather events			Direct impacts — Changing averages	Indirect impacts — Migration	Indirect impacts — Coastal degradation
	Heat waves	Floods	Droughts			
Mortality (cause-specific)						
Drowning		X				
Physical trauma		X				
Heat exhaustion	X					
Fire	X		X			
Suicide		X				
Respiratory diseases						
Asthma				X		
Acute lower respiratory tract infections			X			
Mental diseases						
Depression		X	X		X	
Post-traumatic stress disorder		X	X		X	
Reproductive diseases						
Perinatal complications[a]		X				
Amenorrhea		X				
Rodent- and vector-borne diseases						
Leptospirosis		X				
West Nile virus		X				
Tahyna		X				
Malaria		X		X	X	
Dengue				X		
Tick-borne encephalitis				X		
Lyme borreliosis				X		
Water- and food-borne diseases						
Cholera		X	X	X	X	
Dysentery		X			X	
Hepatitis A		X			X	
Salmonella		X		X		
Acute toxicity						X
Other						
Malnutrition			X		X	
HIV/AIDS					X	
Allergies				X		
Dehydration			X			
Dermatitis						X
Gastroduodenal ulcer disease		X				

Source: Rabie et al. 2008.

Note: x denotes evidence for link between defined exposures and health outcomes, and blank cells denote no evidence of association.

a. Complications include pregnancy loss and disorder (premature delivery, missed abortion, birth asphyxia, premature rupture of membranes, and intrauterine growth retardation).

reliant on subsistence farming tend to resort to permanent climate-driven migration, as in situations where drought has decreased the land area that can be cultivated.

As elsewhere, impacts of climate change in ECA countries come on top of existing conditions and patterns of vulnerability. Migration linked to climate change will add to already high levels of migration, and recipient countries may find their resources are overstretched, particularly in the delivery of health services.

ECA accounts for one-third of the world's total migration (excluding movement of people between industrialized countries), partly because of the high level of migration since the breakup of the Soviet Union. Migration in ECA forms two main streams: first, from Eastern Europe and the former Soviet Union to Western Europe; and second, within the states of the former Soviet Union (Mansoor and Quillin 2007). Russia is the main destination for migrants in the second stream, while the main sending countries in ECA are Albania, Georgia, and Kazakhstan.

People in the region move to pursue economic opportunities, and a challenging or changing climate already does and will continue to influence the economic decision to migrate. A chronic lack of rain pushes families and whole communities to relocate, usually in a depleted and highly vulnerable state. Recurrent drought in Moldova between 1990 and 2007—including a 45-day heat wave in 2007—hit the country's agricultural sector hard as water resources became increasingly scarce. The resulting outflow led to concentrations of Moldovans in large cities abroad, such as Rome and Moscow (IOM 2007).

In Kazakhstan, flooding has caused widespread displacement. Unusually warm days and heavy rains in February 2008 resulted in the inundation of 48 settlements in southern Kazakhstan, forcing 13,000 people from their homes. Most moved into camps or relatives' homes. But in some cases, the floods did long-term damage to farmland and irrigation canals, making restoration of earlier living patterns unlikely. Looking ahead, some anticipate more flooding in the area, possibly displacing as many as 250,000 people (UN OCHA 2008).

Migration can lead to illness and premature death in three ways (see figure 3.1). *First,* dislocated people are stressed and exhausted, without access to safe water and sanitation. This makes migrants more vulnerable to infectious and psychological illnesses, as well as worsening chronic conditions. Adding to this vulnerability, uprooted people have limited access to medical services. Migrants often have little choice but to work as unskilled labor in high-risk, unhealthy jobs. The psychological stresses of culture shock, language barriers,

FIGURE 3.1

How Migration Affects Health and Health Systems

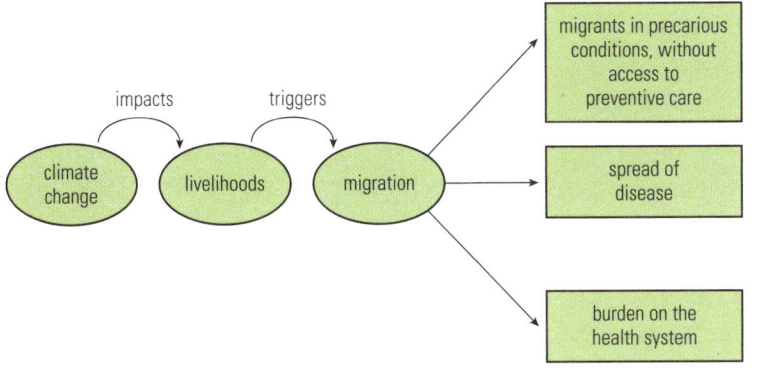

Source: Rabie et al. 2008.

possible discrimination, and overarching insecurity tend to worsen other health conditions.

Second, health systems may be unprepared to deal with the infectious diseases and other illnesses brought by the migrants—either from their home countries or from somewhere along their journey. Infectious diseases are a special challenge for health systems, which must treat individual patients and trace and contain the vectors of disease in order to protect the public.

Migrant populations are vulnerable to type 2 diabetes, cardiovascular diseases, and tuberculosis. According to the European Society for Clinical Microbiology and Infectious Diseases, several large European cities have already experienced tuberculosis epidemics related to increased migration from Africa, Asia, and Latin America. A survey taken in Athens in 2004 and 2005 found positive tuberculin skin tests for 96 out of 1,460 immigrants from Albania, Bulgaria, Romania, the former Soviet Union, Africa, and Southeast Asia (Antypa et al. 2007).

Migrants have higher rates of hepatitis B and C, HIV/AIDS, and malaria than the native populations (Gushulak and MacPherson 2006). A 2005 study within the European Union found that 46 percent of the HIV/AIDS diagnoses were among immigrants, with most of the infections originating outside the European Union (Hamers et. al 2006).

Meanwhile, in Spain a study showed that all of the 24 children admitted to hospitals with malaria between 1997 and 2005 were children of immigrants (Martinez-Baylach et al. 2007). Between 1992 and 2001, Albania reported 114 cases of malaria, all among immigrants who contracted the disease before arriving in Albania (WHO 2002). If the health systems of the destination countries are not prepared to deal with a heavier disease burden, immigrants will not receive proper care, and the uninfected population—both immigrant and native—will be at risk.

Third, immigrants are often underserved by health systems in their destination countries, because of poor communication, restricted access, or discrimination. Lacking access to primary and preventive care, as well as information about available services, immigrants postpone treatment, resorting finally to costly visits to the emergency room. According to evidence from Germany, the greater prevalence of unsafe working conditions and high fertility rates among immigrants also lead to a high number of health complications (Sinn et al. 2001).

Adaptation options for dealing with migration, discussed in more detail in the next section, might include expanding laboratory facilities to screen for previously unfamiliar diseases as well as familiar ones, such as tuberculosis. In addition, countries will need to design communication and education campaigns that help immigrants surmount barriers to health care.

Assessing Vulnerability and Prioritizing Protections

Countries can take stock of their exposure to climate change and its impact on concurrent health problems. One starting point is to collect data that answers some basic questions: What share of the population lives near the coast? What is the history of flash floods? How much of the population is over age 75? How many people are living with diabetes?

In addition, an assessment of exposure and sensitivity must be accompanied by an analysis of adaptive capacity in the health sector. Determinants of adaptive capacity include the following:

- *Economic resources:* public expenditure on health as a percentage of gross domestic product

- *Technology:* adequacy of technological assets in place for responding to health risks

- *Human capital:* the quantity and skills of health professionals, including research specialists

- *Access to risk-spreading mechanisms:* insurance products that enable a society to spread the financial losses associated with the health outcomes occasioned by climate change

- *Access to and ability to manage information:* the availability of critical indicators basic to understanding health risks, including public health surveillance tools, emergency communications, and a system for monitoring changing averages

- *Institutionalized practices:* clinical guidelines, performance assessment protocols, and systems for emergency preparedness

- *Attention to equity:* a measure of how evenly access to and use of health services are distributed throughout the population, and how evenly health deficits are shared.

Adapting health systems to the realities of climate change requires a reliable flow of information and collaboration across organizations. Public health is affected by actions taken in many other sectors. For example, if a country's energy sector increases surge capacity to support cooling during heat waves, heat-related distress and death will decline. Recognizing the probability of flooding and threats to the water system, governments can lower health risks by reducing the vulnerability of water facilities to floods and droughts (see chapter 6).

Health adaptation policies can be grouped into two categories: responsive, which reduces vulnerabilities arising from climate changes that have already occurred; and anticipatory, which addresses health outcomes associated with projected future climate change. Box 3.2 provides a number of adaption actions, both responsive and anticipatory, for government health professionals and the general public to use in response to floods and heat waves, two of the most likely and damaging climate extremes.

In addition to these extremes, climate change involves long-term shifts in average temperatures and precipitation levels, which carry long-term health implications. Governments should strengthen monitoring and surveillance activities in order to detect any new epidemics that might surface. Hygiene should be improved across the board (for example, in food preparation), and vaccination programs and health education increased. A map of high-risk areas should be developed, along with plans for vector-control programs.

Anticipating an increase in migration as a result of climate change, governments should establish screening for tuberculosis and other services for uncommon diseases that might arrive with new residents. Health facilities must inform immigrants of the available health services and perhaps hire more professionals from sending countries. The governments of destination and departure countries could work together to coordinate these actions.

BOX 3.2

Adaptation Strategies for Floods and Heat Waves

Anticipatory strategies for floods

Governments:

- Establish systems to communicate with the public, health professionals, and emergency responders.

- Design education campaigns for populations at risk, including evacuation plans.

- Set up multilingual information systems that can function during and after floods and power outages.

- Divide regions into risk zones based on historical and projected trends for setting investment priorities and informing the public of risks.

- Limit settlement in flood plains with updating and enforcement of zoning laws.

Health institutions and professionals:

- Increase laboratory diagnostic capacity, and strengthen disease-related databases.

- Increase awareness about vector-borne diseases.

- Waterproof facilities, and create safe storage for key equipment.

- Train staff for emergency conditions, including hospital evacuations.

- Back up patient files on computers.

- Create flood-resistant communications systems.

- Create a back-up supply of safe water for hospitals, and invest in purification equipment.

General public:

- Understand safety procedures and priorities in the event of a flood.

- Participate in insurance schemes and other mechanisms for spreading the financial risk.

- Demand a variety of flood-control policies from government.

Responsive strategies for floods

Governments:

- Deliver necessary public awareness materials, and work with media to get key information about the emergency into circulation.

- Ensure public hygiene is maintained.

- Increase levels of human and animal vaccination in year of floods.

- Survey contaminants and environmental threats.

- Ensure access to food, water, and shelter for the most vulnerable persons.

Health institutions and professionals:

- Employ sound surveillance methods to detect and contain epidemics.

- Communicate with government and the public about outbreaks of disease.

- Organize post-flood epidemiological monitoring.

- Include psychological testing to pick up on stress-related factors.

- Provide social support to vulnerable groups.

General public:

- Drink from only safe water supplies, and boil or chlorinate tap water.

- Discard suspect food, remove any dead animals and disinfect contaminated areas, and always wear protective gear.

- Treat furniture and rooms for vector-borne diseases that might come from rodents or insects.

- Clean flooded basements promptly to avoid mosquitoes and molds.

- Use insect repellent.

Anticipatory strategies for heat waves

Governments:

- Ensure the power system has adequate surge capacity.

- Plan future housing to maximize natural ventilation.

- Include space for trees in urban designs.

- Plan back-up water supplies.

- Coordinate forecasting and early warning systems across local authorities.

- Create cool spots and havens using natural and designed systems.

Health institutions and professionals:

- Inform patients of their particular vulnerabilities to heat stress.

- Connect health professionals with forecasting and warning systems.

- Coordinate with government on a public awareness plan, with special outreach to vulnerable groups.

continued

BOX 3.2 *Continued*

- Ensure adequate staffing for emergency periods.

- Create heat wave hotline and Web-based services for public inquiries.

- Design and implement a communication strategy around limiting the effects of smoke and smog.

General public:

- Stay attuned to summer weather forecasts, and know the health risks, including one's own personal medical vulnerabilities associated with extreme heat.

- Agree in advance on possible leave policies from work.

- Advocate for policy makers to adopt heat wave plans.

Responsive strategies for heat waves

Governments:

- Provide continuous electricity during heat waves, with priority for health care facilities.

- Guarantee a flow of public information about government activities, forest fires, and emergency programs.

Health institutions and professionals:

- Monitor health of patients, including outpatient group, particularly elderly and chronically ill persons.

- Ensure that patients understand the seriousness of heat-induced conditions.

- Use media to expand awareness of ways to stay healthy during extreme heat.

General public:

- Avoid strenuous activities and stay indoors during hours of maximum heat.

- Drink a lot of fluids, but avoid alcohol and caffeine.

- Refuel cars at night to lessen gas vapors, and reduce car use.

- Guard against forest fires, and be ready to evacuate if needed.

- Reach out to elderly and vulnerable persons.

Source: Rabie et al. 2008.

CHAPTER 4

Climate Change Will Make Water and Land Management More Complex

Nicola Cenacchi and Marianne Fay

Physical impacts will vary depending on whether climate change manifests itself through slow changes in averages, through more frequent extreme events, or through sudden catastrophic changes (such as a collapse in the North Atlantic current or a collapse of the Greenland or Antarctic ice sheet). Slowly occurring changes are controllable for most human-managed systems—although not always for ecological ones. Extremes are, of course, much harder to cope with and more likely to impose irreversible damages.

This chapter describes the impact of slow-moving averages and more predictable extreme events (or so-called slow-onset disasters such as droughts), concentrating on direct physical impacts. The chapter looks at how climate change might complicate water resource management, reviews how such change is likely to affect coastal areas of Eastern Europe and Central Asia (ECA), and discusses the impact of a receding permafrost line.

This chapter is based on four background papers prepared for this book, "Adaptation to Climate Change in Coastal Areas of the ECA Region" and "Biodiversity Adaptation to Climate Change in the ECA Region" by Nicola Cenacchi; "Expected Impact of the Changing Climate on Russia and Central Asia Countries and Ongoing or Planned Adaptation Efforts and Strategies in Russia and Central Asia Countries" by Alexey Kokorin; and "Climate Change Projections and Impacts in Russian Federation and Central Asia States" by Vladimir Kattsov, Veronika Govorkova, Valentin Meleshko, Tatyana Pavlova, and Igor Shkolnik.

More Difficult Water Resource Management— Too Much or Too Little of a Good Thing?

As seen in chapter 2, ECA will face both more floods and more droughts. Rainfall is expected to become more intense and variable, resulting in increased flood risk everywhere in the region—particularly in Eastern and Central Europe. Droughts will be a serious issue for Central Asia, the Caucasus, and Southeastern Europe. Water management will become more complex everywhere, but—at least in the period up to 2030—this will be mostly driven by natural climate variability, sociodemographic trends, and unsustainable water resource management.

What Climate Change Means for Water Resource Management

Climate change can cause or exacerbate water stress in a variety of ways beyond reduced precipitation. Increased temperature reduces water availability by increasing evaporation while at the same time causing an increase in demand (for irrigation or recreational purposes) and affecting water quality by intensifying the effect of aquatic pollution.

Warmer temperatures also reduce the share of precipitation that falls as snow, which is a natural mechanism for storing water that is gradually released in spring and summer.[1] Climate change affects sea-level rise and storm surges, which can lead to salinization of coastal aquifers (see the discussion of coastal areas in the following section). Finally, more concentrated rainfall and a decline in snowfall, with its water retention function, will likely result in a lower recharge of aquifers as saturated soil conditions lead to more surface runoff.

Increased flooding is a growing concern across all of ECA. Central Europe has been particularly affected in recent years—especially Bulgaria and Romania. The flooding, which can be riverine or coastal, can also come from rising underground water; this is a serious issue for a number of Russian Federation cities, such as St. Petersburg. Increased intensity of rainfall along with storms and more rapid snowmelt in the spring are the climatic drivers of floods.

A number of nonclimatic factors already threaten the sustainability of water resources, including urban growth, changing land use, and unsustainable agricultural and industrial water use (Arnell and Delaney 2006; Holman et al. 2005). One study (Vörösmarty et al. 2000) shows that for the early part of this century, water stress in Europe and Asia will be almost entirely driven by increased water demand linked to socioeconomic developments. Similarly, evidence suggests that floods are often linked to poor land use and river basin management.

Generally, climate-related changes to freshwater systems have been small, compared with such nonclimatic drivers as pollution, regulation of river flows, wetland drainage, reduction in streamflow, and lowering of the groundwater table (mainly due to extraction for irrigation). This mosaic of stresses calls for a shift toward more sustainable practices before the impacts of climate change are more strongly felt over the next 20 years.

A Varying Regional Picture, but More Flooding

As discussed in chapter 2, Central and Southeastern Europe, Central Asia, and the Caucasus will experience reduced precipitation, while the rest of the region (Northern Europe; Russia, except for the Northern Caucasus; and Kazakhstan) will mostly see increased rainfall. Annual runoff—the water that runs over land, and a measure of water availability—is projected to decrease in Central and Southeastern Europe and Central Asia but increase for most of Russia and the Baltics (maps 2.4 and 4.1). Winter runoff is projected to increase, especially in European Russia, with the most pronounced changes projected for the spring for the rest of ECA (Kattsov et al. 2008).

Projections are grim in terms of frequency and intensity of extreme flooding in ECA. The Danube and Tisza valleys in Hungary are very prone to frequent flooding. Floods are projected to be more frequent

MAP 4.1

Changes in Annual River Runoff by 2041–60 Relative to 1900–70

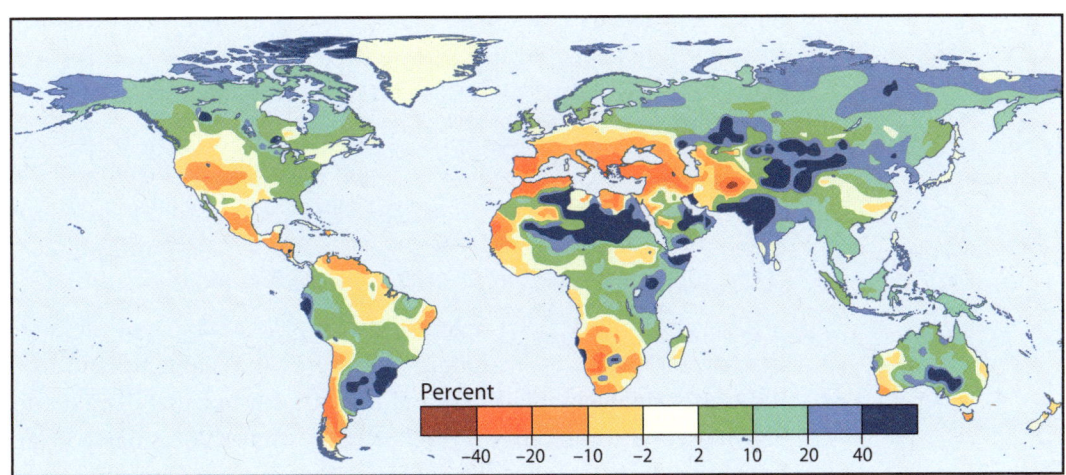

Source: Carter et al. 2007: 184.

in Northern, Central, and Eastern Europe, as well as in Asian Russia. Intense short-term precipitation and the risk of flash flooding will rise across most of Europe. Flood protection traditionally relies on reservoirs in highland areas and dykes in lowland areas.[2] However, other planned adaptation options are becoming more popular, such as expanding zoned floodplain areas (Helms et al. 2002), emergency flood reservoirs (Somlyódy 2002), preserved areas for flood water (Silander et al. 2006), and flood warning systems, especially for flash floods.

Anticipating and responding to flood risk will require intelligently managed institutions that identify water use trends, areas vulnerable to climate change, and opportunities to respond to the emerging challenges. Particular measures for flood management include effluent disposal strategies under conditions of lower self-purification in warmer water; design of water and wastewater treatment plants to work more efficiently, even during extreme climatic conditions; and ways of reusing and recycling water (Luketina and Bender 2002; Environment Canada 2004; Patrinos and Bamzai 2005).

Climate Change Will Compound Central Asia's Serious Water Shortages[3]

Central Asian countries are confronting a shared problem of future water shortages, probably the most dramatic in the region. Increased winter precipitation will be more than offset by declining summer precipitation and warmer temperatures. Declines in river runoff—estimated to be about 20 percent in the next 50 years—will compound already unsustainable water management.

The nature and extent of water vulnerability varies. In Kazakhstan, the decline in runoff is expected to be milder, but there is a potential problem of water resource management in the Ili River basin, which is shared with China. The Kyrgyz Republic and Tajikistan will have enough water for their own needs but may not be able to meet demand in their role as critical suppliers of water to the region—the Kyrgyz Republic has 30 percent and Tajikistan 41 percent of total water resources for the five countries (table 4.1).

The rapid melting of the glaciers of the Kyrgyz Republic and Tajikistan is worrisome, particularly in Tajikistan, whose glaciers contribute 10 to 20 percent of the runoff of the major river systems of the region (up to 70 percent during the dry season). The glaciers are critical to the Amu-Darya water basin, the most important in Central Asia and the principal source of water for Turkmenistan. In addition, the Kyrgyz

TABLE 4.1

Water Resources of Central Asia: Suppliers of the Main Rivers

Country	Amu-Darya River basin (km³/y)	Syr-Darya River basin (km³/y)	Balkhash Lake basin (km³/y)	Issyk-Kul Lake basin (km³/y)	Tarim River basin (km³/y)	TOTAL (km³/y)	Share of country's resources in regional total (%)
Kazakhstan	—	2	24	—	—	**26**	**17**
Kyrgyz Republic	2	28	0.3	4	7	**41**	**27**
Tajikistan	63	1	—	—	1	**65**	**43**
Turkmenistan	3[a]	—	—	—	—	**3**	**2**
Uzbekistan	5	6	—	—	—	**11**	**7**
TOTAL	**79[b]**	**37**	**24**	**4**	**8**	**151[b]**	**100[b]**
Share of basin's contribution to regional total (percent)	52	25	16	2	5	100	

Source: Alamanov et al. 2006: 105, as quoted in Kokorin 2008.

Note: km³/y = kilometers cubed per year, a measure of annual volume.

a. Including Iran's small contribution in Amu-Darya runoff.

b. Including Afghanistan's contribution in Amu-Darya runoff; that is, 6.2 km³/y or 4 percent of the total water resources of the region.

Republic is also seeing a troubling decline, partly attributable to climate change, of the water level of Lake Issyk-Kul, which is important to the country's economy and ecosystems.

The water situation in Turkmenistan and Uzbekistan is dramatic, but it would be so even without climate change. Uzbekistan is the main water consumer of the region—it is the most populated country with an economy largely based on irrigated farming. Almost all (90 percent) of its water resources come from mountains located in other countries. Adaptation will require more sustainable use of water, starting with implementation of low-water-consuming technologies and more effective irrigation management. It may also include reservoirs and regulation of runoff.

Unsustainable water management has caused the Aral Sea to shrink, which will be made worse by climate change. Once the fourth largest lake in the world, the Aral Sea is nearing extinction, having decreased over the last four decades from 68,000 square kilometers (km²) to about 28,000 km² (Glantz and Zonn 2005). Once 178 species inhabited the Aral region; now fewer than 40 do so (Alamanov et al. 2006). Salt air pollution from the open sea bottom is dangerous for agriculture as well as human and animal health. Warming temperatures are only making it worse—for example, by increasing evaporation over the 1,300-km, man-made Karakum channel.

Significant damage has already occurred in the Amu-Darya River Delta, and measures to manage today's stresses will be even more important to refine as the climate changes. The Amu-Darya is a key source of water for Tajikistan, Turkmenistan, and Uzbekistan, which share in its use and, therefore, will have to coordinate efforts to save the Aral. The government of Uzbekistan is attempting to stabilize the sea with a program that includes development of buffer protection basins, which are chains of local water reservoirs in the Amu-Darya River Delta and surrounding areas.

Many problems, however, have not yet been addressed, including modernization of archaic and wasteful irrigation systems and other climate-sensitive infrastructures. Although most of the countries involved have developed adaptation policies, implementation is slow (Kokorin 2008). And, while integrated river basin management is essential throughout the region, it is complicated by the transboundary nature of the region's water resources (box 4.1).

More Stress on Already Stressed Coastal Areas[4]

Coastal areas, defined as "areas on and above the continental shelf ...; areas routinely inundated by saltwater; and adjacent land, within 100 km from the shoreline" (Martinez et al. 2007), are subject to impacts from both the sea and the land. This exposes them to the influence of climate change either directly (sea-level rise, storm surges, floods, and droughts), or indirectly through events that originate offsite but whose consequences propagate down to the coasts (such as river floods and changes in the seasonality, pulses, and quality of runoff from inland sources).

Coastal vulnerability varies tremendously across ECA's four basins (the Baltic Sea, the East Adriatic Balkan coast and Mediterranean coast of Turkey, the Black Sea, and the Caspian Sea) and the Russian Arctic Ocean. Some basins are experiencing a decrease in sea levels (Caspian and northern Baltic), while others face varying degrees of sea-level rise. Seawater acidification caused by higher concentrations of carbon dioxide and increases in water temperature affects them all. Vulnerability in all basins is exacerbated by poor coastal management and existing stresses—pollution; overfishing; construction too close to the coast; and the damming of rivers, which prevents sediment flows from reaching the coast, worsening erosion.

Vulnerability also depends on whether a significant share of a country's population or economic activity is situated in low-elevation coastal zones. This share is highest in Latvia, where 34 per-

---

BOX 4.1

Placing More Emphasis on River Basin Management

Integrated Water Resource Management (IWRM) is a systematic approach to planning and management that considers a range of supply-side and demand-side processes and actions, and incorporates stakeholder participation in decision processes. It identifies and balances trade-offs among the water management objectives of environmental sustainability, economic efficiency, and social equity. IWRM simultaneously addresses the two distinct systems that shape the water management landscape. The biophysical system—including climate, topography, land cover, surface water hydrology, groundwater hydrology, soils, water quality, and ecosystems—determines the availability of water and its movement through a river basin. Factors related to the socioeconomic system, driven largely by human demand for water, shape how available water is stored, allocated, and delivered within or across river basin boundaries.

Integrated analysis of the natural and managed systems is arguably the most useful approach to evaluate management alternatives. This type of analysis uses hydrologic modeling tools that simulate physical processes, including precipitation, evapotranspiration, runoff, and infiltration. In managed systems, analysts must also account for the operation of hydraulic structures (such as dams and diversions), as well as institutional factors that govern the allocation of water between competing demands, including consumptive demand (such as agriculture) and nonconsumptive demands (such as hydropower generation).

At the river basin level, IWRM seeks to manage the sharing of costs, benefits, and impacts among all uses and users across a river basin. But it is also the most challenging approach to water resource management because of the obstacles created by sector and administrative boundaries.

Source: Contributed by Shelley McMillan.

---

cent of the population lives in coastal zones less than 10 meters (m) above sea level, and is significant in a number of other ECA countries (table 4.2).

Baltic Sea

Variations in the Baltic Sea level are strongly affected by the uplift of the Scandinavian plate in the north and the lowering of the southern Baltic coasts. This relationship, combined with the increase in mean ocean level, has resulted in a recorded sea-level rise of 1.7 millimeters per year in the southeastern Baltic, but a decrease of 9.4 millimeters per year in the northern part (HELCOM 2007). Projected sea-level

TABLE 4.2
Share of the Population Living in Low-Elevation Coastal Zones
(Less than 10 m above sea level)

Country	Total population in low-lying coastal zone	As a share of national population (%)
Latvia	814,288	33.6
Albania	317,894	10.1
Georgia	328,396	6.2
Lithuania	186,901	5.1
Turkey	2,449,027	3.7
Romania	760,789	3.4
Croatia	139,930	3.0
Ukraine	1,315,903	2.7
Poland	973,501	2.5
Russian Federation	3,552,274	2.4
Moldova	87,726	2.0
Bulgaria	121,581	1.5
Montenegro	8,583	1.3
Bosnia and Herzegovina	700	0.0

Sources: SEDAC, CEISIN.
Note: Armenia, Azerbaijan, Belarus, Hungary, Kazakhstan, Kyrgyz Republic, FYR Macedonia, Serbia, Slovak Republic, Tajikistan, Turkmenistan, and Uzbekistan have no exposed population.

rise will depend mostly on land uplift and global sea-level rise, with the latter apt to balance the former in the northern areas. The best studies on coastal vulnerability in the Baltic have been conducted in Estonia and Poland, discussed below.

To date, no obvious trend of sea-level rise has been recorded in Estonia, whose coast is only moderately vulnerable. A 1 m sea-level rise, for example, would threaten important ecological sites but few settlements (Kont et al. 2008) because the coast is sparsely populated. The two primary vulnerable sites are the capital city of Tallinn and the Sillamae industrial center; the latter is the dumping site for the radioactive wastes of a former uranium enrichment plant. These wastes, which regularly leach into the soil and water, are separated from the sea only by a narrow dam.

Increased storminess and sea level rise could cause radioactive material to be flushed directly into the Baltic. The city of Tallinn is protected for one-third of its coastline by seawalls, but the defense system will require adjustments because of increased storminess.

Increased coastal development—partly for tourism—would increase vulnerability.

Poland's coast seems more vulnerable. General circulation models project increased frequency and strength of storm conditions along with a continued rise in sea level that could reach 45–65 centimeters by 2100 (Pruszak and Zawadzka 2008). Poland's low-lying and mostly sandy coasts are exposed to flooding and erosion, which has increased since the 1970s because of the rise in sea level, greater storminess, and sediment starvation brought about by the regimentation of rivers.[5] The socioeconomic vulnerability of Poland's coast is particularly high at the eastern and western extremities (the cities of Gdańsk, Gdynia, and Szczecin). Sensitivity could increase as coastal development, which began in the 1990s after a jump in gross domestic product (GDP), continues its course.

Runoff into the entire Baltic Sea will likely increase over this century as precipitation linked to climate change becomes heavier, altering the delicate coastal water nutrient balance. More runoff will translate into a greater input of nutrients and possibly will intensify eutrophication (HELCOM 2007).[6] This development, combined with the projected continued warming of sea water, could spark increased phytoplankton growth that could be harmful to human and animal health (see chapter 3).

Caspian Sea

The Caspian Sea has displayed significant sea-level fluctuations. The causes, which are not well understood, may include changes in precipitation and runoff, along with tectonic movements.

Climate models project a 6 m *decrease* in the level of the Caspian Sea from its1975 level by the end of the twenty-first century because of increased surface evaporation, which is expected to exceed the augmented runoff from the Volga (Elguindi and Giorgi 2007; Renssen et al. 2007). A significant drop in sea level combined with increasing temperatures will impact fish stocks and put additional stress over the already imperiled sturgeon population. The drop in sea level would also affect infrastructure and economic activity, increasing costs for industry (mainly oil and gas) and transport.

Unfortunately, awareness of the unpredictable sea levels has not discouraged coastal development on land freed by the retreating sea. Past rises have caused vast damage along sections of the Caspian coastline; the Russian coast is a prominent example (Frolov 2000; GEF 2002). A new drop in level could result in another rush to occupy newly available land, exposing the population to potentially dangerous substances, such as pesticides, arsenic, and other heavy metals, locked in coastal sediments.

Mediterranean Sea (East Adriatic and Turkey's Mediterranean Coast)

The Mediterranean is a difficult place to gather data on sea-level forecasts. Tectonic activity, changes in density of deep waters, and local changes in air pressure systems complicate measurement activities (Karaca and Nicholls 2008). Within the East Adriatic, observations of sea-level rise at different locations show great differences, with the average rising at one site and dropping at another.

In Croatia, for example, studies project a significant sea-level rise, but with high levels of uncertainty (e.g., +65 ± 35 cm by 2100). For this reason, a joint project of the United Nations Development Programme and the Global Environmental Facility is working on a qualitative assessment of vulnerability to a wide range of possible sea-level changes (Bari, Grbec, and Bogner 2008).

Croatia's rocky coast would protect it well against a small sea-level rise (for example, 20 cm) but not against much higher rises. Particularly vulnerable are tourism, fisheries, and shipping infrastructures built right up to the shore. Further analysis is needed to understand the vulnerabilities of coastal cities, notably to saltwater intrusion into groundwater tables (Bari, Grbec, and Bogner 2008).[7]

The northern part of Albania is highly sensitive to floods and more frequent storms. Unregulated urban development has allowed building up to the shoreline, exposing infrastructures to a high risk of weather-related damages. The impact will vary with the extent of sea level rise: the 48–60 cm rise projected for 2100 would flood coastal areas and cause significant saltwater infiltration (UNDP-Albania 2002), whereas the 20–24 cm projected for 2050 should not have major impacts despite the fact that all the coasts of Albania are considered lowlands. Along Turkey's Mediterranean coast, the coastline's geophysical characteristics imply low physical vulnerability to sea-level rise, but some settlements and productive activities will be vulnerable (UNDP-Turkey 2007).

Nevertheless, sea-level rise and storm surges could impact tourism and agriculture (Karaca and Nicholls 2008). The delta plains of Gediz, Seyahn, and Ceyhan where land has been reclaimed for agricultural use are especially vulnerable (Karaca and Nicholls 2008). The movement of populations toward coastal cities is amplifying the sensitivity of the socioeconomic system to sea-level rise. Istanbul is particularly exposed, as 10 percent of the population lives within 1 km of the shore, and the city by itself accounts for 21 percent of national GDP. The biggest concerns involve saltwater intrusion, particularly in two coastal lagoons and in Lake Terkos, which supply fresh water for the city (Karaca and Nicholls 2008).

Black Sea

Sea-level rise has been higher in the Black Sea than in the Mediterranean (27 ± 2.5 millimeters per year, versus 7 ± 1.5 millimeters per year; Valiela 2006), although the few studies that exist lack consistency. The Georgian coast appears to be subsiding relative to the rest of the Black Sea basin (Karaca and Nicholls 2008), while the Russian coast with its numerous ports and high economic activity will be vulnerable to floods and saltwater intrusion into freshwater aquifers (Frolov 2000). Ukraine is already experiencing erosion problems that are prompting a loss of housing, arable land, and industrial and touristic sites.

The Black Sea coast of Turkey is vulnerable mainly in a few deltaic areas (Karaca and Nicholls 2008). Storm surges are already affecting some settlements, and worsening conditions may bring damages to the 23 ports along the Black Sea. Furthermore, storms, erosion and sustained flooding are predicted to damage the very important east-west road system that runs along the coast quite near to the shoreline.

Climate change will add to the stresses already felt in the Black Sea coast. The economically critical fishing industry—already threatened by overfishing and pollution—will be further stressed by the projected increase in water temperatures. The Black Sea is also an important source, refinement point, and transport route for oil and gas, and there are fears that increased storminess and erosion will stress oil and gas infrastructure on the Georgian, Russian, and Ukrainian coasts. Accidents would spread further pollution. Coastal landfills in the Black Sea are pollution hotspots (GEF 2007); and along coasts such as Georgia's, sea-level rise and coastal erosion may further damage these landfills and increase the volume of pollutants flushed into the sea.

Finally, the damming and channeling of rivers, along with poorly managed coastal development, is altering the sediment balance and distribution, resulting in erosion problems. In Russia, Georgia, and Ukraine, unregulated building close to the shore is also advancing erosion and increasing sensitivity to climate impacts.

Declining Arctic Ice, Tundra, and Permafrost[8]

Climate impact is fastest and most visible in the Arctic region. Projections of Arctic Ocean ice show decreases in area and mass throughout the twenty-first century, with the decreases more pronounced in the seasonal minima (September) than in the seasonal maxima (March) (Kattsov et al. 2007, 2008). While there is a lot of variation among models, studies project a mean reduction in September ice on the order of 40 percent by mid-century. Zhang and Walsh (2006) project the year-round ice to decrease on the order of 45 percent to 65 percent in the last two decades of the twenty-first century, while seasonal ice (meaning ice that melts in the summer) is projected to increase 14 percent to 28 percent over the same period. In some models, the Arctic Ocean's ice cover becomes entirely seasonal by the end of the century.

Regarding Russian permafrost, seasonal thaw depths are projected to increase by more than 50 percent in the most northern parts of Siberia by 2050, and by 30 percent to 50 percent in most other permafrost areas (Anisimov and Reneva 2006). Projections also suggest increased seasonal thawing depths along with a northward shifting of the boundary between seasonal thawing and seasonal freezing (map 4.2). Finally, over the next 100 years, Russia's tundra is projected to shift to forest (Scholze et al. 2006), with estimates of total converted tundra area ranging from 10 percent to 50 percent (Anisimov and Vaughan 2007).

The implications of the large scale thawing of the permafrost go far beyond the urgent biodiversity problem caused by the loss of ice in the Arctic and the impacts on buildings and infrastructure (chapter 6). Permafrost is thought to hold about twice the amount of carbon in the atmosphere. While some of it would be captured by the encroachment of trees in the tundra, emissions of carbon as carbon dioxide or methane—a much more potent greenhouse gas—from microbial decomposition of organic carbon in thawing permafrost could amount to roughly half those resulting from global land-use change during this century (Schuur et al. 2008). The large-scale thawing of the permafrost would be a major catastrophic event that could lead to runaway global warming.

Threats to Biodiversity Are Significant[9]

ECA countries are home to a significant part of the world's biodiversity. The region includes the world's largest contiguous steppe and intact forest ecosystems along with 21 mountain chains (Brylski and

Shifting Boundaries and Degradation of Permafrost by Mid-Century

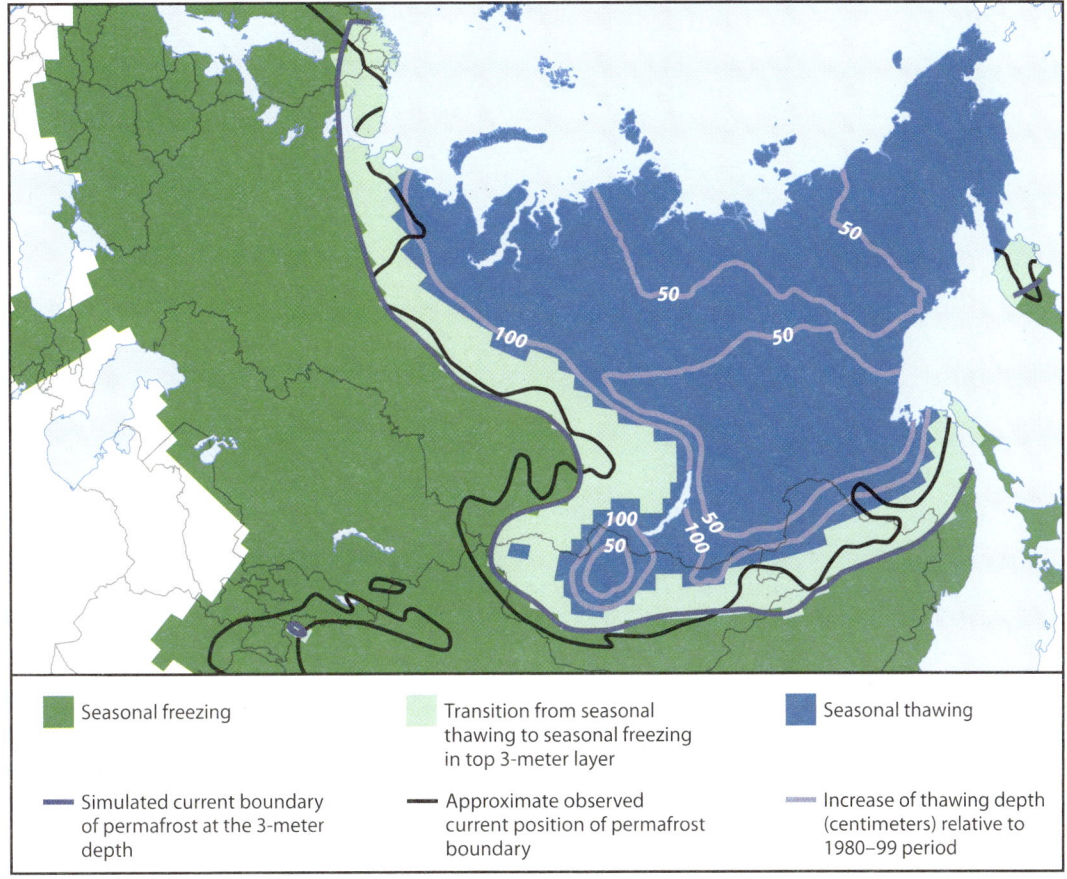

Seasonal freezing

Transition from seasonal thawing to seasonal freezing in top 3-meter layer

Seasonal thawing

— Simulated current boundary of permafrost at the 3-meter depth

— Approximate observed current position of permafrost boundary

— Increase of thawing depth (centimeters) relative to 1980–99 period

Source: Kattsov et al. 2008.

Abdulin 2003), 9 of 15 major biomes, and nearly 100 different ecoregions (map 4.3). Much of this biodiversity is already threatened. Indeed ECA is also home to 26 of the World Wide Fund for Nature (WWF) Global 200 priority areas (map 4.4), and three hotspot regions: the Mediterranean basin, the Caucasus, and the Mountains of Central Asia.[10]

The first impact of increasing temperatures will be to change species' ranges, meaning it will induce a movement of ecosystems themselves. Some species and ecosystems—those that already occupy the most extreme areas in the mountainous or arctic regions—will have nowhere to go and will face possible extinction. Others will not be able to adapt fast enough, given the unprecedented rates of temperature change. As species push northward or upward, warmer and wetter conditions are also expected to create more opportunities for invasive species to expand their range

MAP 4.3
Major Biomes and Ecoregions in ECA

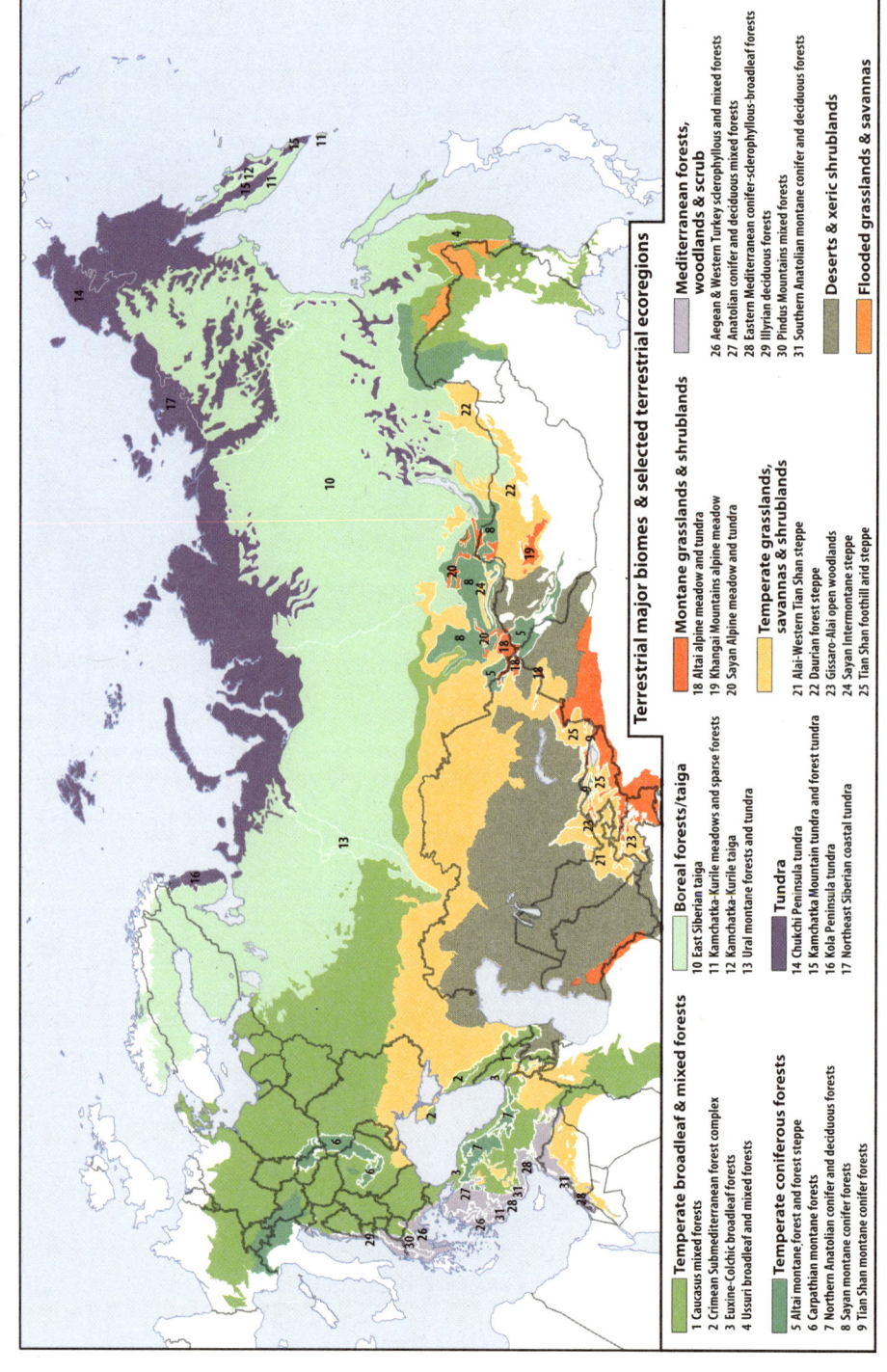

Source: World Bank map using WWF data.

Note: The ECA region contains nearly 100 terrestrial major biomes and ecoregions. This map only includes labels of the areas discussed in detail in table 4.4. A map with a full listing of numbered areas is available in Cenacchi (2008b).

MAP 4.4
WWF Global 200 Priority Areas

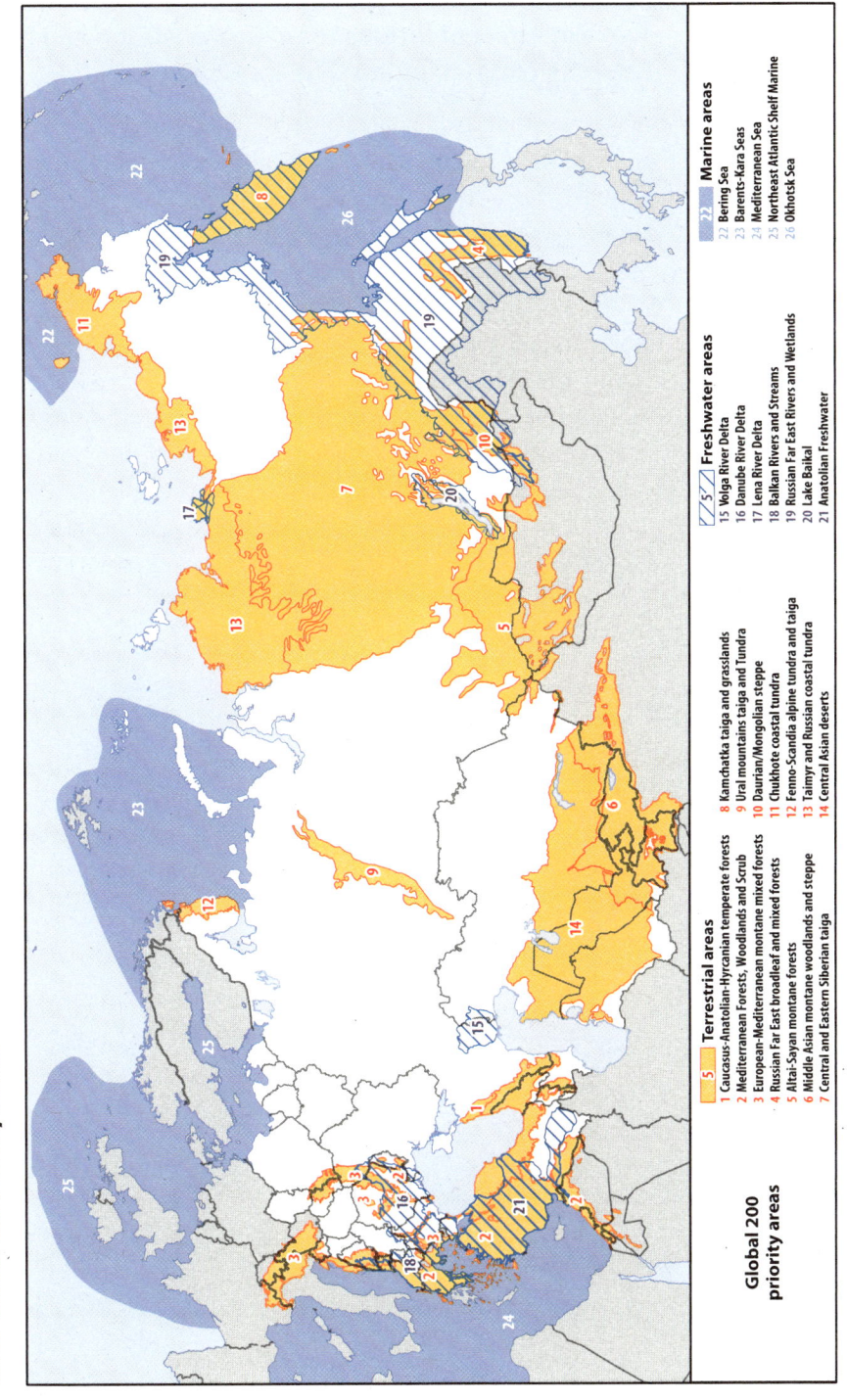

Global 200 priority areas

Terrestrial areas

1 Caucasus-Anatolian-Hyrcanian temperate forests
2 Mediterranean Forests, Woodlands and Scrub
3 European-Mediterranean montane mixed forests
4 Russian Far East broadleaf and mixed forests
5 Altai-Sayan montane forests
6 Middle Asian woodlands and steppe
7 Central and Eastern Siberian taiga
8 Kamchatka taiga and grasslands
9 Ural mountains taiga and Tundra
10 Daurian/Mongolian steppe
11 Chukhote coastal tundra
12 Fenno-Scandia alpine tundra and taiga
13 Taimyr and Russian coastal tundra
14 Central Asian deserts

Freshwater areas

15 Volga River Delta
16 Danube River Delta
17 Lena River Delta
18 Balkan Rivers and Streams
19 Russian Far East Rivers and Wetlands
20 Lake Baikal
21 Anatolian Freshwater

Marine areas

22 Bering Sea
23 Barents-Kara Seas
24 Mediterranean Sea
25 Northeast Atlantic Shelf Marine
26 Okhotsk Sea

Source: World Bank map using WWF data.

(Alcamo et al. 2007, Reid 2006). Climate change will also affect the timing of natural cycles, such as flowering or mating seasons.[11]

Two Key Lines of Intervention: Conservation and Minimizing Stresses Not Related to Climate Change

A first step entails tackling directly those stressors that undermine adaptation of species and ecosystems—one of the arguments behind establishing protected areas. However, protected areas—still basic to biodiversity conservation—will become less effective as habitat ranges, and with them the distribution of species, shift.

The key to an adaptation strategy then is an anticipatory framework enabling the natural systems to adapt on their own, to the extent possible, to climate change. The preferred approach is the establishment of networks of protected areas, shielded by buffer zones and connected through vegetation corridors, that allow species' migration along altitude and latitude gradients (Price and Neville 2003).[12] But to be effective, they must have a landscape-regional (or "bioregional") approach (box 4.2).

BOX 4.2

Bioclimatic Models

Ideally, the design of protected areas should be informed by bioclimatic modeling, that is, modeling of the range shifts of species. Regional modeling of biodiversity responses, including magnitude and direction of change, is necessary because global models are not useful for conservation of biodiversity (Hannah et al. 2002).

One of the challenges is how to test the various models' predictive ability; obviously, there are no future data to test the predicted distribution of species in relation to climate change. One solution has been to make use of past climate and species distribution data (Araujo and Rahbek 2006). But this type of data is hard to find, and testing of models is restricted to a few regions and a few species within them.

In ECA, there is a large amount of untapped historic data that may be extremely useful (one example is "Chronicles of Nature," an official document produced in each of the approximately 200 protected areas of Russia, recording past changes in the distribution of species, both flora and fauna). The situation calls for a program to track and recover such material and use it to support regional biomodeling future changes.

Source: Cenacchi 2008b.

One example is the central conservation initiative of the European Union (EU), the Natura 2000 Network—26,000 protected areas covering all member states with a total area of 850,000 km^2 or more than 20 percent of EU territory. It does not exclude all human activity; rather, it includes both nature reserves and privately owned land where extensive agriculture or pasture are allowed and managed according to sustainable practices (European Commission 2005).

The UNESCO World Network of Biosphere Reserves is an example of the extension of the landscape approach to the global scale. Unlike national protected areas, this network spans national boundaries. Examples within ECA are the biosphere reserves at the southwestern end of the Tien Shan Mountains and at the Carpathians.

Adaptations by Biome

ECA's hotspots and WWF's Global 200 priority areas—spanning a number of ecoregions—face the intense stresses of habitat destruction and fragmentation, requiring forward-looking adaptation measures (table 4.3).

TABLE 4.3
Categories of Adaptation Options

Protected areas	• Identify ecosystems, species, and processes particularly sensitive to climate change. • Design areas to protect species, habitat, and ecosystems. • Evaluate and improve the management and monitoring capabilities.
Conservation networks	• Develop a network of protected areas endowed with buffer zones and connected through corridors that allow species to move along different altitudes and latitudes. • Implement stepping stones and landscape management to allow movement through mostly anthropogenic landscapes.
Bioregional approaches	• Establish a network of protected areas covering and crossing political boundaries (such as the EU Natura 2000 Network) to allow more protection of species' movement and to preserve functions of large ecosystems.
Participation in management	• Involve local people in the management of protected areas. • Improve locals' livelihoods by decreasing their dependence on natural resources.
Monitoring	• Implement as a key element of any adaptive management activities. • Use GLORIA—Global Observation Research Initiative in Alpine environments—a long-term observation network to detect effects of climate change.
Supporting policies	• Design and implement policies and plans for specific geographical areas and for sectors and agencies, including legal provision and economic instruments.
Minimizing of stresses not related to climate change	• This is a landscape-level prescription and applies also to protected areas: minimize pollution; control exotic species; and minimize pressures from land-use changes, development, and tourism.

Source: Cenacchi 2008b.

Grasslands and Forests

Grasslands and forests are vulnerable due to the increased risk of wildfires and invasive species. A key adaptation strategy centers on control of exotic species. Monitoring the migration of wild grazers is also critical. The Daurian steppe (table 4.4), which contains rare plant species, is currently exposed to unregulated road construction and unsustainable grazing practices. Both factors are potentially disastrous for maintaining resiliency to climate change—as plant genetic diversity will be key in efforts to identify forage and other plants that thrive in the changing climate—and have to be addressed urgently as an initial adaptation measure.

Given the need for northward migration, physical barriers to migration must be avoided. For instance, in the north of the Central Asian steppe lies a vast swath of agricultural land that is difficult for species to cross. Corridors or "stepping stones" could allow the southern grassland species to move into and across the land occupied and used by people.[13]

Mountains

Global warming will cause mountain ecosystems to shift upward (rather than northward). This will result in a loss of the ecological zones at the summit of the mountains—since species have no place left to which they could migrate (Price and Neville 2003). This phenomenon is already observable all over the world, from the Italian Alps to the Urals to the Altai-Sayan Mountains. In mountain chains with a north-south geographical orientation (such as the Urals), the permanent ecosystem disruption may be delayed as species may find temporary refuge in the northernmost areas.

In the Urals, the main threats are the clear-cutting of old-growth forests, mining, agriculture and pasture, air pollution, and tourism. However, the threats are not equally distributed along the chain. While the mountain tundra seems to have been degraded all across the ecoregion (apart from a few protected areas), the northern taiga is still in relatively good condition.[14] Its protection is therefore critical. In the Altai-Sayan and Khangai mountains, stressors are hunting, poaching, logging, overgrazing, and mining. In the Carpathians, poaching and air and water pollution are the main issues, along with logging for ski resorts and building of hydroelectric dams.

TABLE 4.4

Biomes, Areas of High Conservation Interest, and Adaptation Measures

Biome	Global 200 Priority Areas and Conservation International Hotspots in ECA	Anticipatory planned measures to promote autonomous adaptation
Alpine or montane ecosystems (temperate coniferous forests; montane grasslands and shrublands)	• Carpathian montane forests (6) • Altai-Sayan montane coniferous forests (5, 8)[a] • Altai-Sayan alpine meadows and tundra (18, 20)[a] • Khangai Mountains alpine meadow (19) • Tien Shan montane conifer forests (9)[b] • Ural montane forests and tundra (13)	• Minimize all non-climate-related threats (habitat destruction and fragmentation, pollution, and so forth). • Promote the establishment of protected areas and protected networks. • Promote the participation of local people in conservation by improving their livelihoods. • Monitor and actively control the introduction and spread of exotic species.
Temperate broadleaf, mixed, or coniferous forests	• Caucasus-Anatolian-Hyrcanian temperate forests (1, 2, 3, 7)[c] • Ussuri broadleaf and mixed forests (4) (Russian Far East broadleaf and mixed forests priority area)	• Control current threats, particularly degradation, fragmentation, and exotic species. • Modify protected areas to take climate change–induced shifts into consideration, and to increase connectivity. • Change management of forests to larger biogeographic scales, including increased control over buffer zones. • Make sure all habitat types are represented in the protected areas, and protect mature and old growth stands.
Boreal forests or taiga	• East Siberian taiga (10) (Central and Eastern Siberian taiga priority area) • Kamchatka taiga (11, 12)	
Mediterranean forests, woodlands, and shrub	• East Adriatic coast, Greece, Turkey, and East Mediterranean, South Anatolian coasts (26–31)[d]	
Temperate grasslands and steppe	• Sayan intermontane steppe (24)[a] • Alai-Western Tien Shan steppe (21)[b] • Gissaro-Alai open woodlands (23)[b] • Tien Shan foothill arid steppe (25)[b] • Daurian forest steppe (22)	• Monitor and control the spread of exotic species through roads. • Regulate unsustainable grazing (e.g., in the Daurian steppe). • Promote connectivity to prevent fragmentation during migration processes.
Arctic ecosystems (including tundra)	• Kamchatka mountain and forest tundra (15) • Chukchi peninsula tundra (14) • Kola peninsula tundra (16) (Fennoscandia alpine tundra and taiga) • Northeast Siberian coastal tundra (17) (Taimyr and Russian coastal tundra)	• Protect habitat. • Reduce non-climatic stresses (pollution, overharvesting). • Monitor and regulate tourism. • Monitor and control invasive species. • Implement the WWF "Conservation First" principle.
Freshwater areas	• Volga River Delta • Danube River Delta • Lena River Delta • Balkan rivers and streams • Russian Far East rivers and wetlands • Lake Baikal	• Protect a variety of potential habitats, including thermal refugia. • Protect water flow and hydrological characteristics. • Protect habitat connectivity between rivers, lakes, and wetlands. • Control spread of exotic species.

Source: Cenacchi 2008b.

Note: The names of the priority areas are supplemented with numbers identifying the relevant ecoregion in map 4.3.

a. = part of the Altai-Sayan priority area.

b. = part of the Middle Asian montane woodlands and steppe priority area (also a Conservation International Hotspot).

c. = a Conservation International Hotspot.

d. = part of the Mediterranean Basin hotspot.

As a priority, conservation networks (ideally, collaborations of governments, nongovernmental organizations, and technical experts) must be recognized by neighboring countries to eliminate political obstacles. For example, the Altai-Sayan mountain environments are shared by Kazakhstan, Mongolia, and Russia; the Carpathians span across the Czech Republic, Poland, Romania, the Slovak Republic, and Ukraine. Finally, because poverty is endemic in these areas, conservation goals are unlikely to be achieved unless local livelihoods are improved and dependence on unsustainable exploitation of natural resources is reduced.

Arctic

Given the scale of projected climatic impacts over the Arctic, the only adaptation strategy is to enhance natural autonomous adaptation capacity. This requires tackling current stressors, particularly pollution. The city of Norilsk is one of the major sources of sulfur in the world because of its nickel smelting plants. Sulfur dioxide has already destroyed a vast part of the forests in the Taimyr and central Siberian tundra—one of the WWF priority areas (National Geographic Society 2001). The Lena River Delta, one of WWF's freshwater priority areas, is partially protected, but the delta is threatened by mining activities, forestry, and agriculture development (WWF 2008). These activities make the delta even more vulnerable to the increased coastal erosion provoked by the combination of permafrost melting and sea-level rise. The developed areas around the Lena wetlands represent a barrier to species migration, in addition to causing a coastal squeeze, impeding the retreat of wetlands in the face of sea-level rise.

Notes

1. The melting of mountain snowpack over the summer is a natural mechanism for redistributing precipitation across seasons. Normally, greater winter precipitation is stored as snow and ice, and then gradually released throughout the spring and summer as temperatures rise.
2. The discussion of flood protection is adapted from Bates et al. (2008).
3. This section is based Kokorin (2008), a background paper produced for this book.
4. This section is based on Cenacchi (2008a), a background paper produced for this book.
5. Subsidence has had little effect on the Polish coast, being only of 1 millimeter a year.

6. *Eutrophication* literally means *overnourishment*. The term refers either to atypical algal blooms or to the massive death of organisms following the decomposition of algae and the loss of oxygen in the water. These events are triggered by the availability of enormous quantities of both inorganic and organic nutrients, such as from runoff from fertilized fields.

7. At the same time that the sea is rising, projected declines in precipitation and increased extraction would lower the level of underground freshwater supplies, making inflow of saltwater even more likely for any given amount of sea-level rise.

8. The material on projections is based on the background papers by Westphal (2008) and Kattsov et al. (2008), while the discussion of the implications of permafrost melting and adaptation options is from the background paper by Kokorin (2008).

9. This section is based on Cenacchi (2008b), a background paper commissioned for this book.

10. The WWF Global 200 priority areas are a set of ecoregions where conservation efforts and resources should be concentrated—based on the level of species richness and endemism. Hotspots are areas "featuring exceptional concentrations of endemic species and experiencing exceptional loss of habitat" (Myers et al. 2000).

11. These cycles are known as *phenological* cycles. The effect of a phenological shift on a species depends on whether the other species on which it relies—for food, pollination, or seed dispersal—change with it.

12. A *buffer zone* has the double purpose of benefiting local populations while providing an additional level of protection to the conservation area; it is intended for both conservation and development fostering research, tourism, and so forth, and for prohibition of activities such as logging, mining, and construction. *Corridors* typically indicate landscape vegetational structures that facilitate the migration of both animal and vegetal species, as well as the exchange of human populations, to reduce the chance of genetic isolation.

13. In corridors, stepping stones are smaller disconnected areas or protected habitats that have been tested to facilitate movement of animals, including insects, birds, and large mammals.

14. http://www.worldwildlife.org/wildworld/profiles/terrestrialpa.html.

CHAPTER 5

The Unbuilt Environment: Agriculture and Forestry

William R. Sutton, Rachel I. Block,
and Jitendra P. Srivastava

For the managed environment of Eastern Europe and Central Asia (ECA)—farms, commercially exploited forests, and fisheries—climate change is already happening. Moldova's drought-stricken agricultural sector and Central Europe's forest fires during the 2003 heat wave provide a harbinger of the challenges the farming and forestry sectors will face over the coming years (Fink et al. 2004).

However, the impact of climate change will vary across ECA countries; some areas and sectors are likely to experience significant new stresses, while others might see positive impacts. There are also variations as to when, and how directly, different areas and sectors must cope with climate change impacts. The increased frequency of heat stress, drought, and flooding caused by climate change threaten to reduce crop yields and livestock productivity in many areas. Shorter and less harsh winters may result in potential productivity gains in others. In the forestry sector, increased risks of fires and pest outbreaks will negatively affect the health of forests (Easterling et al. 2007).

Yet the region's comparative inefficiency and low productivity in agriculture and forestry far outweigh the benefits and risks of climate change (Olesen and Bindi 2002). The recent global food crisis

This chapter is based on "Adaptation to Climate Change in Europe and Central Asia Agriculture" by William R. Sutton, Rachel I. Block, and Jitendra P. Srivastava, a background paper prepared for this book.

revealed the inability of a number of ECA countries to respond to increased prices and demand, raising concerns about skewed incentives and the region's capacity to adapt to the challenging shifts projected under climate change scenarios. To change this, ECA's leaders, farmers, and foresters will need to address the productivity gaps with Western Europe in both agriculture and forestry.

Adaptation is essential to protect and enhance rural livelihoods in ECA. Farms, forests, and fisheries play a crucial role in rural poverty reduction, employment, economic growth, and food security. But adaptation is also critical to respond to increased demand for food as global population soars, and to offset the declines in yields that many countries outside of ECA will experience as a result of climate change. Indeed, the ECA countries that stand to benefit from moderate temperature increases (1°C to 3°C in the global annual average) will play a vital role in meeting the world's growing demand for food.

Agriculture and forestry can also help mitigate further climate change and may offer opportunities for tapping into carbon finance. Globally, agricultural production and deforestation account for up to 30 percent of greenhouse gas emissions, second only to the power sector (IPCC 2007c). These sectors therefore offer opportunities for carbon sequestration, such as afforestation or minimum tillage agriculture. But mitigation strategies do not protect societies against the climate change impacts already in evidence, or those in the pipeline as a result of past greenhouse gas emissions.

All governments in the region will need strategies that allow their countries to take advantage of potential gains from climate change, as well as minimize risks and threatened losses. A country or subregion may have the potential to expand farm outputs under certain climate change scenarios, but this potential will not be realized if the infrastructure is crumbling and binding institutional or market barriers aren't removed.

Despite variations across countries, all face challenges in adapting to climate change and increasing efficiency and sustainability. In Central Asia, the unforgiving topography and hydrology would make adaptation difficult, even if institutions were functioning at optimal effectiveness. Southeastern Europe, home to some of the most productive land in the region, is projected to suffer from drought, heat waves, and more frequent forest fires. In the north, there are potential benefits from climate change, but these will be achieved only if countries adjust institutional frameworks to support new patterns of production. Even then, other barriers will persist, including the poor quality of soils in the northern Russian Federation, the lack of public services and infrastructure, possible social dislocations, and local environmental damage (Dronin and Kirilenko 2008).

This chapter reviews the impacts of climate change on farming and forestry in ECA, highlighting the region's inherent sensitivity to climate change and limited adaptive capacity, and what these imply for both winning and losing regions and sectors. It concludes with recommendations about possible adaptation measures.

First, however, we discuss the continued importance of agriculture in many countries in the region—particularly for poverty concerns.

Climate Impacts Will Exacerbate ECA's Persistent Problem of Rural Poverty

Despite the perception of ECA as an urbanized region, agriculture remains an important part of the livelihoods of many, especially the poor (Alam et al. 2005). Agriculture is a particularly significant share of gross domestic product (GDP) in Central Asia, Southeastern Europe, and the South Caucasus (table 5.1).

Across ECA, roughly one-third to one-half of the population lives in rural areas, with the figure approaching two-thirds in Central Asia. Even in Kazakhstan and Central and Eastern Europe, a significant share of the population remains rural, despite the fact that

TABLE 5.1

Agriculture Matters: Poverty and the Rural Economy in ECA

Region	Agriculture as share of GDP (%)	Rural population (percentage of total population)	Rural poverty rate (percentage of rural population)	Share of poor residing in rural areas (%)	Rural extreme poverty rate (percentage of rural population)	Share of extremely poor residing in rural areas (%)
Southeastern Europe	12.3	35.4	61 with Turkey, 44 without	45	20 with Turkey, 9 without	46
Central and Eastern Europe	8.7	36.1	44	48	10	54
Baltics	5.3	35.2	33	42	3	39
Russian Federation	5.6	27.1	53	34	14	42
South Caucasus	12.0	45.9	80	48	30	49
Kazakhstan	6.7	42.2	79	52	31	64
Central Asia	27.0	64.1	94	69	62	73

Sources: World Bank 2008; Alam et al. 2005.

Note: Data on agriculture and rural population are from 2006 where available; otherwise, 2005. Poverty line = US$4.30 per person per day. Extreme poverty line = US$2.15 or less per person per day. Both poverty lines use purchasing-power parity dollars, 2002, 2003, or 2004 if available. For rural poverty, Central and Eastern Europe is Moldova, Romania, and Ukraine; Central Asia is the Kyrgyz Republic, Tajikistan, and Uzbekistan; Southeastern Europe is Albania, Bosnia and Herzegovina, Bulgaria, the Former Yugoslav Republic of Macedonia, Montenegro, and Serbia, with and without Turkey.

agriculture accounts for a smaller portion of their economies. In much of the region, one half or more of poor people live in rural areas, with three-fourths of extremely poor people in Central Asia living in the countryside. Thus, any poverty strategy must take into account new stresses felt in rural areas as a consequence of global climate change.

Forests, though not as significant economically as agriculture, remain important for rural livelihoods both through direct employment and through ecosystem services (such as the provision of wood and food, or protection against erosion and floods). Forestry accounts for only about 0.1 percent of GDP in much of Central Asia and the Caucasus; however, outside market systems, forest resources may be significant to rural communities, particularly with respect to the fuelwood. The market importance of forestry is somewhat higher in Central, Eastern, and Southeastern Europe.[1]

Rural poverty rates in ECA are significantly higher than national average rates, and the share of rural people in poverty ranges from a low of one-third in the Baltics subregion to a staggering 94 percent in Central Asia. Across the region, about half of poor people are found in rural areas (but the share is one-third in Russia and two-thirds in Central Asia). Thus, most ECA countries other than Russia have a poverty profile heavily influenced by conditions in rural areas, particularly with respect to agriculture.

Agriculture is uniquely effective in reducing poverty.[2] The inverse, of course, is that setbacks in agriculture—whether losses or missed opportunities—are disproportionately damaging to the rural poor. Thus, even if climate change has only a small impact on the overall economy, it could have a profound effect on the portion of the population living below the poverty line, or the entire population of a particular district or locality. And at the level of the household or individual, negative shocks could have a more persistent impact on welfare through effects on health and nutrition (Randolph et al. 2007).

Livestock activities are important to many vulnerable groups in the ECA region and may be undergoing structural shifts as the demand for meat, eggs, and dairy products increases in Asia's fast-growing economies. The delicate balancing of grain allocation between staple food and animal feed may become more difficult in the context of changing global demand. Shocks from climate change could add to an already uncertain mix of factors, potentially exacerbating the current global food and feed crisis (Sirohi and Michaelowa 2007). Untangling the interplay of shifting global demand, climate change, and patterns in livestock-related land use—and teasing out the policy implications—is a continuing endeavor worldwide.

Models Predict There Will Be Winners and Losers in ECA

Beyond the undisputed conclusion that climate change will add to the vulnerability of most, if not all, of the rural populations already living in poverty, the effects of changing weather patterns on ECA's agriculture and forestry will be hugely varied. Climate and agro-economic information, while far from comprehensive, provides sufficient data to illustrate the scope of climate shifts already underway, along with some future changes and their potential impacts. Yet uncertainty remains, notably as to how private interests or institutions might respond to the new opportunities and risks that come with warmer, wetter, or drier weather.

Observed Climate Changes and Impacts

Changes in climate and their impacts on agricultural systems and rural economies are already evident throughout ECA. The growing season has lengthened in locations stretching from Germany to European Russia (Maracchi, Sirotenko, and Bindi 2005). Chapter 2 noted that extreme events have occurred with greater frequency and intensity in Europe, most recently in the 2003 summer heat wave over much of the continent, and more intense flooding in Central and Southeastern Europe. A decline in precipitation along the northeastern coast of the Mediterranean has caused significant drought-related damages in the agricultural economies of Southeastern Europe (Alcamo et al. 2007). Drought-induced economic losses in all sectors have been calculated for the region and are in some cases large.[3] Further, successive weather extremes can amplify existing stresses: Moldova's resilience was already weakened by past storms and droughts when a major drought occurred in 2007, bringing greater economic disruption.

In Central Asia, Kazakhstan, Asian Russia, and the Arctic, twentieth-century increases in temperature have surpassed the worldwide warming average, reaching as much as 3°C (Cruz et al. 2007; Kattsov et al. 2007). The frequency and intensity of extreme events have increased, including heat waves, extreme cold days and winter storms, heavy rains and floods, and droughts (Alcamo et al. 2007; Cruz et al. 2007).

In the mountainous South Caucasus, observed changes have exhibited geographic variations in both direction and magnitude; while average temperature has increased slightly and average precipitation declined slightly, localized impacts have been larger (Hovsepyan and Melkonyan 2007). Severe droughts have become increasingly common in the North and South Caucasus and Central

Asia, worsened by poor land management, soil degradation, and reduced rain or runoff (World Bank 2005).

Projected Impacts: The Agronomic View

ECA as a whole, as well as individual ECA countries, is characterized by a tremendous variety of climates. It encompasses both warm, dry regions where agriculture and forests are projected to experience significant damage from climate change, and colder zones where agriculture and forestry could benefit from warmer temperatures and increased precipitation (see table 5.2 for a summary of changes in agricultural potential; detailed regional information on impacts is in box 5.1). Small-holder farms in Albania that depend on irrigation may be hard hit by droughts and heat waves, while in parts of Poland, a longer growing season and warmer winters may allow greater crop diversity

TABLE 5.2

Crop Potential in the ECA Region Today and Possible Shifts by 2100

General climate class	Average temperature of warmest months (°C)	Crop-growing period (days)	Crop potential	ECA regions in 2008	ECA regions in 2100
Very cold	8.5–11	<90	Quick-maturing green root vegetables (e.g., lettuce and radishes)	Parts of Arctic Region, Siberia, and Far East (Russia)	
Cold	10.5–16	<100	Early varieties of vegetables (e.g., cabbage, spinach, turnips, early varieties of barley, oats, buckwheat, flax, hardiest local varieties of apples and pears)	Northern parts of Urals, Western Siberia, and Far East	
Moderately cold	15–20	100–50	Winter wheat; spring wheat; rye; barley; oats; legumes; flax; potatoes; cabbage; beets; locally adapted winter-hardy varieties of apples, pears, plums	Baltics, northern parts of Central Russia and Volga Region and Southern Siberia, Northern Kazakhstan	
Moderate	18–25	150–80	Wheat, corn, rice, sunflower, soybeans, melons, early cotton, vegetables, walnuts, peaches, apricots, apples, grapes, cherries, plums	Ukraine, southern parts of Central Russia and Volga Region, Northern Caucasus, Central Europe	
Warm	>25	>180	Cotton, citrus, figs, grapes, olives, wheat, corn, rice, vegetables during winter, subtropical perennials (e.g., tea, nuts, and a variety of fruit crops)	Central Asia, Caucasus, Southeastern Europe, Turkey, Southern Kazakhstan	Compare to South Mediterranean and Middle East

Sources: Olesen and Bindi 2002; Maracchi, Sirotenko, and Bindi 2005; Lampietti et al. 2009; Parry et al. 2007; European Commission 2007; Alexandrov 1997; Sirotenko, Abashina, and Pavlova 1997; Hovsepyan and Melkonyan 2007.

BOX 5.1

Estimated Agronomic Impacts of Climate Change in ECA to 2050: A Summary

Southeastern Europe, including Turkey

Increased variability in yields of cereals and other crops[a,b]

Decreased precipitation in all seasons, yet more storms and floods

- soil erosion from wind, storms, and floods[a]
- increased evapotranspiration, soil salinization
- increased irrigation demand, stress on water supply
- especially severe water stress in southern Turkey

Higher average temperature, very hot summers, heat waves, and droughts

- faster maturation, shorter development period, with water shortage and heat stress, grain sterility, lower yields of many cereals, oilseeds, and pulses (i.e., determinant crops)[a]
- decreased yield or quality of onions,[c] cool-weather vegetables[a]
- longer season for warm-weather vegetables
- possible shifts to higher altitude of some crops (especially mountainous Turkey)
- increased variability of grape quality, quantity, and vulnerability to pests, but potential benefit from CO_2 fertilization (see note below)
- expansion of drought-tolerant olive, citrus, and fig[a,b]
- but tree crops highly vulnerable to storms, pests[c]
- winter survival and subsequent proliferation of pests[d]

Livestock

- heat stress and both indigenous and nonindigenous disease in livestock threaten milk and meat production[b,c]
- heat, water scarcity decrease forage production leading to shortage in late summer[b]

Central and Eastern Europe

Right on the line between north (wetter, milder winter) and south (drier, hotter winter), so not yet clear if climate and, thus, impacts will be similar to the neighbors to the north or to the

continued

BOX 5.1 *continued*

south. Potential yield increases mostly shown in Alps and Carpathians,[e] where significant agriculture not actually feasible. Disagreement among sources, including range from benefits in some places to large losses around the Black Sea (Eastern Romania, Moldova, Southern Ukraine—hot and dry), with little agreement for all of Ukraine.[c,e]

Increased storms, but ambiguous magnitude and direction of precipitation change

- tree crops vulnerable to storms

- even if no change in region-wide average precipitation, possible yield decline if too wet in the north (see Baltics) or even slightly drier in the south (see Southeastern Europe)

Same amount of warming in winter and summer

- faster maturation, shorter development period, which may lower yield of many cereals, oilseeds, and pulses (determinant crops)[a]

- potential for northward expansion of warm weather crops like oilseeds, pulses, vegetables[c]

- potatoes more variable, possibly limited by low soil moisture[a]

- winter survival and subsequent proliferation of pests

- too warm and dry for rain-fed cereals in parts, but suitable for more tree crops, including fruit and nuts, more natural pasture biomass for animals

- possible increase in area of winter wheat and rye

Baltics

Increased variability in yields of cereals and other crops.[f,g] Potential yield gains require more fertilizer and pesticides.[c] No consensus on strongly positive nor strongly negative yield projections overall; generally small, positive for initial moderate warming, becoming unpredictable and possibly negative as mean temperature increases further.[d,e]

Increased precipitation, floods

- risk of soil erosion

- excess soil moisture limits days suitable for machinery use[a]

- spring planting disrupted by April or May rains

- harvest disrupted, damage from water-logging, or molding of harvested grain if excess rain in autumn[a]

Milder winters and higher average temperature

- faster maturation, shorter grain-filling period, lower yield of winter wheat,[a] but now possible to use higher-yielding spring wheat

- potential for northward expansion of warm weather crops such as oilseeds, pulses, and vegetables[a]

- either no changes or favorable changes in potato and sugar-beet yields, but increased variability[a]

- winter survival and subsequent proliferation of pests[a]

- more varieties of apples, plums, and pears

Livestock

- increased survival, reduced winter feed requirements for livestock[c]

- forage, grassland may benefit but only with proper drainage[d,e]

Russia: Baltic and Western Arctic

Large change, especially in Arctic, and, thus, large uncertainty.

Marked increase of precipitation, especially in winter, and of surface water

- risk of soil erosion and nutrient leaching from excess rain

- excess soil moisture limits days suitable for machinery use[a]

- spring planting disrupted by April or May rains

- harvest disrupted, damage from water-logging, or molding of harvested grain if excess rain in autumn[a]

Much milder winters and higher average temperature

- potential for northward expansion of temperate cereals, vegetables, and pulses in Baltic, and of hardiest crops into uncultivated land[c]

- longer growing season[g]

- potato yields more variable, though with average increase[c]

- expansion of leaf-bearing forest and steppe range into current tundra and taiga[c]

- change in composition of forests, and possible increase in value for timber production

Livestock

- increased survival, reduced winter feed requirements for livestock[c]

- forage and grassland may benefit but only with proper drainage[c,d]

continued

BOX 5.1 *continued*

Russia: Central and Volga

Increased variability in yields of cereals and other crops.[f,g]

Small increase of precipitation, mostly in winter, and of surface water

- given small increase, unclear if there will be sufficient moisture, given temperature increases, in some months

- extremely low runoff (drought) events threaten output[d]

Much milder winters and hotter summers, higher average temperature

- potential for northward expansion of winter cereals and crops such as oilseeds, pulses, and vegetables, as well as fruit crops currently grown in North Caucasus[c]

- longer growing season

- winter survival and subsequent proliferation of pests[d]

Livestock

- increased survival and reduced feed requirements for livestock in winter[c]

- possible heat stress, drying up of grassland in summer[c,d,g]

- possible expansion and intensification of indigenous and nonindigenous disease[d]

- in southern part, productivity of grassland to decline, will need to shift northward

- lower grass production, heat stress, dry summers lead to reduced milk, greater vulnerability to disease[d]

Russia: North Caucasus

Decreased precipitation in all seasons, yet more storms, floods, and soil erosion. Higher average temperature, very hot summers, heat waves, and droughts.

Very similar changes, on average, to South Caucasus, though even higher heat-wave risk. See agronomic impact information for South Caucasus. The area with the greatest potential damages within Russia, given current agricultural importance and nature of projected changes. Plant and animal diseases to occur more frequently.

Russia: Urals and Western Siberia, South Siberia, Siberia and Far East

Marked increase of precipitation, especially in winter, and of surface water; high flood risk

- excess precipitation may limit expansion of cereals otherwise possible from temperature increase alone

- risk of soil erosion

- excess soil moisture limits days suitable for machinery use[a]

- spring planting disrupted by April or May rains

- harvest disrupted, damage from water-logging, or molding of harvested grain if excess rain in autumn[a]

Much milder winters and higher average temperature

- shift of agro-ecological zones on a diagonal gradient toward the northeast, so currently forested or uncultivated land warm enough for cereals and short season vegetables

- expansion of cereals would entail major changes in land use over time

- expansion of leaf-bearing forest and steppe range into current tundra and taiga[c]

- change in composition of forests, and possible increase in value for timber production

Livestock

- increased survival, reduced winter feed requirements for livestock[c]

- forage and grassland may benefit but only with proper drainage[c,d]

Note: South Siberia has a different climatic and agricultural baseline, though projected climate *changes* are similar to the rest of Asian Russia.

South Caucasus

Decrease in surface water; droughts and floods; decline in spring and summer precipitation; small increase on sea coasts in winter

- high risk of summer droughts

- salinization, desertification, and soil degradation[h]

- yield declines for cereals, vegetables, and potatoes from water shortage and excess heat in many areas

- widespread crop failures during droughts

- strain on water supply for irrigated agriculture[h]

Especially hotter in summer, also milder winters

- despite milder winters, more crop-destroying frosts (tree crops, fruits) because of absence of heat-retaining humidity[h]

- longer growing season may allow multiple harvests[h]

- expanded area for cultivation of warm-weather tree crops (figs, nuts) in plains, and expanded area for vegetables (tomato, peppers) and cool-weather tree crops (apples) at high altitudes, but limited by steepness and risk of increased erosion[h]

- potential yield increase and geographic expansion for hot-weather perennials like grapevine, olive, and citrus, but with risk of high variability[c,h]

continued

BOX 5.1 *continued*

- tree crops vulnerable to storms and pests[c]

- winter survival and subsequent proliferation of pests[d]

Livestock

- increased heat stress and disease, but less stress from cold in winter[c]

- outcomes for forage and grassland not clear[h]

Kazakhstan

More rainfall and surface water year-round in north, with very dry summers in south

- despite CO_2 fertilization, increased heat and water shortage cause decline in yields of cotton, rice, fodder, vegetables, and fruit production in irrigated south[d]

- potential expansion of grazing land northward and in formerly virgin marginal lands, ploughed for wheat cultivation

Much warmer throughout year, slightly more in summer

- potential increase in cereal, legume, and oilseed production in cooler, wetter north

- increased fodder production

- increased water demand of plants and drying of soils in warmer months because of higher temperatures, causing drought risk and water scarcity to persist or worsen

Livestock

- initial warming good for livestock, provided sufficient water availability, but after first few degrees, increased heat stress and disease[d]

Central Asia

Unchanged or increased winter rainfall; decrease in rainfall and surface water in spring, summer, and fall; earlier and faster glacial melt, with droughts

- major stress on water resources for irrigation

- decline in cereal yield from water shortage from spring to fall, and from thermal stress[d]

- drought, desertification, soil erosion, salinization

- widespread crop failures during droughts

- increased suitability for drought-resistant tree crops

Hotter summer, milder winter

- greater water demand for rice production with higher temperatures[d]

- despite CO_2 fertilization, increased heat and significant water shortage cause decline in cotton yields[d]

Livestock

- marginal grasslands at risk for aridization, desertification

- heat stress reduces milk production

Sources: a. Olesen and Bindi 2002; b. Lampietti et al. 2009; c. Maracchi, Sirotenko, and Bindi 2005; d. Parry et al. 2007; e. European Commission 2007; f. Alexandrov 1997; g. Sirotenko, Abashina, and Pavlova 1997; h. Hovsepyan and Melkonyan 2007.

Note: Carbon fertilization refers to an expected increase in yield of many crops as the result of higher ambient carbon dioxide (CO_2) concentrations, because CO_2 is an input into photosynthesis, and more CO_2 means more photosynthesis and, thus, growth, and because higher concentrations can reduce respiration (that is, water loss from the "pores" in leaves), thereby increasing water use efficiency. There is still debate about the magnitude of the CO_2 fertilization effect, so estimates with and without it are considered here (Cline 2007).

and increased productivity. Large countries, such as Kazakhstan, which incorporate various climate zones, will be home to both winners and losers as climate change impacts play out. Parts of the country projected to see increasing rainfall could see expanding opportunities for rain-fed, high-yielding winter wheat, while other parts face reduced water availability, sporadic drought, and lower cotton yields.

Ideally, countries could embark on a smooth adaptation process (as illustrated by the arrows in table 5.2), with cereal cultivation shifting northward in Russia and Kazakhstan, and longer growing seasons allowing for increased diversification into high-yield or high-value crops in the cool, temperate areas of Central and Eastern Europe and European Russia. But of course, it takes planning, investment, and effective knowledge services to take advantage of climate-induced opportunities.

Further south, hotter, drier summers pose new risks, with more frequent, intense droughts in Southeastern Europe and Turkey, the North and South Caucasus, and Central Asia. The net effect could well be new limits on output and far greater volatility in crop yields from year to year. In fact, as illustrated in the last column of table 5.2, the model for agriculture in the already warm, dry areas of ECA eventually will be drawn less from local experience than from current practices in the Middle East and North Africa (MENA). Today's management challenges and conflicts over water in MENA offer a sobering picture of what some in ECA must adapt to.

The projected increase in weather extremes presents challenges for agriculture across all parts of ECA. Inundating rains in Russia and the Baltics may interrupt sowing and harvesting of cereals. Storms in Central and Southeastern Europe could destroy tree crops. Alternating drought and intense rain and snowmelt could cause erosion and landslides in the densely cultivated slopes of the Caucasus. Drought combined with the scarcity of irrigation water could accelerate soil degradation; and as vegetation withers, local climate feedback effects could result in less precipitation and worsening drought. Climate change will only worsen the long-term spiral of intensifying aridity in Central Asia and the Southern Caucasus (Alcamo et al. 2007; Lampietti et al. 2009; Cruz et al. 2007; Easterling et al. 2007; Hovsepyan and Melkonyan 2007; Maracchi, Sirotenko, and Bindi 2005; Olesen and Bindi 2002).

Livestock production, also sensitive to weather patterns, could benefit in the north from increased forage production, lower feed requirements, and reduced threat of extreme cold. But in the warmer, drier areas, changing rainfall patterns and extreme heat will affect livestock both directly—through heat stress, lack of drinking water, and changed reproductive patterns—and indirectly—through reduced forage and feed yields. The unwelcome arrival of infectious diseases (such as brucellosis or rabies) because of warmer temperatures, would further stress herds.

Livestock production can add to the climate change problem through overgrazing and local climate feedback effects and, globally, through methane emissions. Livestock activities now contribute 80 percent of all agricultural greenhouse gas emissions (FAO 2006a). If producers respond to declines in the productivity of livestock by enlarging their herds, the result could be overgrazing, pasture degradation, and erosion of watershed catchments, causing devastating local climate feedback effects (Kokorin 2008). This scenario of grasslands becoming dry and barren is already a concern in water-scarce Central Asia, where many people depend on traditional agro-pastoral grazing systems.

ECA's forests face tree loss and degradation from extreme events and from the combination of earlier snowmelt and hot, dry summers. Regional droughts and shifting wind patterns have already increased the frequency and intensity of wildfires, notably in Bosnia and Herzegovina, Croatia, and Serbia in 2007, and in Russia, where approximately 20 million hectares were lost to fires in 2003 alone. Strong winds, which are projected to increase as climate changes, can not only spread wildfires but also spark the initial conflagration. Many suspect that strong winds near electrical wires were the culprit in the 2008 fires in the Turkish province of Antalya, where, in addition to taking lives and destroying trees, the fires devastated vast stretches of productive farmlands.

A changing climate can redistribute tree species, with warming causing shifts to higher latitudes. The new patterns can also provoke outbreaks of insect infestations, as seen in the northern march of damaging pests in boreal forests around the world (Easterling et al. 2007). Similarly, a changed climate sets the stage for the invasion of non-native, harmful plant species into already disrupted forest eco-systems. Plant and pest species will move to higher altitudes in response to global warming, a trend already observed in the expanded range of birch (*Betula pubescens*) into the tundra of Sweden over the last half of the twentieth century.

Projected Impacts: the Economic Models

The Model Estimates

The economic effects of climate change on agriculture include direct yield impacts, which are the most easily estimated, as well as ripple effects across sectors and markets. We take the initial shock to potential crop yields as our starting point before subsequently considering market forces and feedbacks, with particular focus on the international food market. Based on our analysis and estimates available in global synthesis studies, primarily Cline (2007), which are discussed further in box 5.2, we have also attempted to identify potential winners and losers in agricultural output markets.

The results show the potential for the following changes in the agricultural economies of the region:

- net losses in Southeastern Europe and Turkey, the North and South Caucasus, and Central Asia

- gains in the Baltics and in the Urals, Siberia, Far East, and Baltic and Western Arctic subregions of Russia

- mixed or uncertain outcomes in Central and Eastern Europe, Kazakhstan, and the Central and Volga region of Russia.

The subregional summaries (table 5.3) are not meant to be defini-tive because uncertainties remain, but they can help identify poten-tial conditions that farmers and policy makers can shape and respond to on the basis of current knowledge of climate change. Although precise impacts cannot be gauged, a pattern does emerge in which

BOX 5.2

Economic Agricultural Impact Models and Their Limitations

The Cline (2007) estimates have been chosen here because they incorporate both main types of models, agronomic and Ricardian, to arrive at consensus estimates. (For further discussion, see Sutton, Block, and Srivastava 2008.) However, there are a number of reasons to interpret the results with caution. Six major limitations of economic impact models are as follows:

• the lack of ECA-specific data, particularly important in mountainous, glacier-fed, and water-constrained areas, in the initial design of the models

• the reliance on averages to determine yields, when in fact variability, extremes, and nonlinear tipping points may be equally or more important

• oversimplification of hydrology and, thus, failure to consider realistic constraints on water availability

• a partial equilibrium view of resource allocation and production, that is, omission of trade-offs in the allocation of land and water and of market feedback effects

• the lack of consideration of the barriers to adaptation, from the geographic, technological, and infrastructural to the institutional, informational, and financial

• highly optimistic assumptions about a positive supply response from ECA in the face of global shifts in food production potential, demand, and prices, which would require currently absent complementary institutions and investments.

southern areas, already water-stressed, will be vulnerable to the projected higher temperatures and lower precipitation, while higher latitudes could benefit from improved conditions for agriculture. The economic impact models for forests are less developed.

Interpretation and Caveats

At first glance, the impacts on ECA's agricultural economies appear manageable, particularly when compared to those of South Asia or the Sahel, where yields are projected to decrease by more than 25 percent. Because the models all have weaknesses, and because country-level economic projections are rudimentary, decision makers should see the projections as broadly indicative. To date, very little analytical work has been done at the country level in ECA to estimate the economic costs and benefits of climate change impacts and

TABLE 5.3

ECA's Potential Winners and Losers in Agriculture from Climate Change

Region	Based on box 5.1 and Cline (2007)	Yield impacts in 2080s without CO_2 fertilization (%)	Yield impacts in 2080s with CO_2 fertilization (%)
South Caucasus	Likely loser	−17.0	−5.0
Central Asia	Likely loser	−9.0	+4.6
Southeastern Europe and Turkey	Likely loser	Europe: −8.6 Turkey: −16.2	Europe: +5.1 Turkey: −3.6
Central and Eastern Europe	Mixed or indeterminate	−5.0	+8.5
Kazakhstan	Mixed or indeterminate	+11.4	+28.1
Russia: North Caucasus	Likely loser		
Russia: Central and Volga	Mixed or indeterminate		
Russia: Baltic	Potential winner		
Russia: Western Arctic	Potential winner	−7.7	+6.2
Russia: South Siberia	Potential winner		
Russia: Urals and West Siberia	Potential winner		
Russia: Siberia and Far East	Potential winner		
Baltics	Potential winner	−5.0 to +5.0	+9.5 to +27.9

Source: Sutton, Block, and Srivastava 2008.

Note: Relative to the other parts of ECA, Kazakhstan's yield increases are probably an overestimate; more details are provided in box 5.2. CO_2 = carbon dioxide. For an explanation of carbon fertilization, see note to box 5.1.

agricultural adaptation, and even less to address the intra-country distributional implications of climate change.

The estimates are limited in the sense that they can include only trends, but not all climate changes follow a simple trajectory. A key example of the complexity of climate change is the impact on the Syr Darya and Amu Darya rivers, which draw on mountain snowmelt in the spring and early summer and glacial melt in late summer; these rivers provide much of the water for Central Asian farms before eventually draining into the Aral Sea in western Kazakhstan and Uzbekistan. The glaciers of the Tien Shan Mountains of Northern China and the Kyrgyz Republic, a critical source of water, have declined sharply in the past 50 years, with an accelerated retreat in the past two decades (Niederer et al. 2008). As warming continues and winter snowfall is replaced by rainfall, river flow will increase in the winter but decline in the summer when it is most needed. This is because there will be little accumulated snow. Up to 2050, water from the rapidly melting glaciers will increase substantially: estimated increases range from 30 percent to 300 percent (Agaltseva 2008, Cruz et al. 2007). But after these few decades, the flow from the diminished glaciers will slow to the point where Central Asian farms will not have enough water for irrigation. As a result, the Aral Sea will likely shrink further, possibly reversing recent successes in restoring the water level and local eco-systems (Savoskul et al. 2003).

The State and Sensitivities of ECA's Agriculture Today

For any region, the capacity to manage climate change will depend on its demonstrated ability to address a broader range of problems related to the environment and natural resource base. The institutional and economic conditions of countries will shape the ways that countries respond to the challenges posed by shifting weather patterns. Stakeholders engaged in adaptation assessments and planning will need to understand how land is used, and which population groups are vulnerable, as well as the diversity of agricultural practices. A depiction of the region's land use categories appears in map 5.1, while ECA's agricultural characteristics appear in table 5.4.

MAP 5.1
Current Agricultural and Other Land Use in ECA

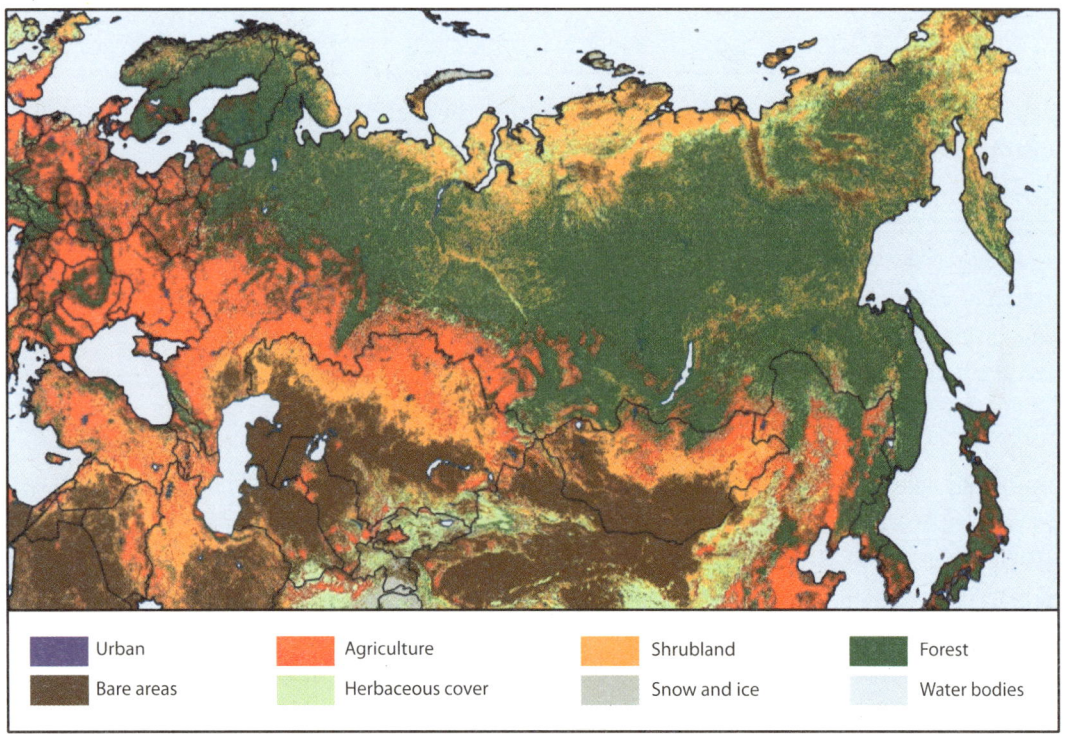

| Urban | Agriculture | Shrubland | Forest |
| Bare areas | Herbaceous cover | Snow and ice | Water bodies |

Source: European Commission 2000.
Note: The map is based on spot vegetation data collected at 1 kilometer intervals.

TABLE 5.4

Characteristics of Current Agricultural Production in ECA

Region	Distribution, ownership, and productivity of agricultural land	Major crops and products	Cropland irrigation and water supply
Southeastern Europe, including Turkey	Farms of Bulgaria now privatized; Croatian and Macedonian farms privately owned. Albania, Serbia, and Montenegro mostly private but unclear ownership rights, and some remaining inefficient collectives. Limited efficiency from excessive fragmentation of holdings throughout region. In Turkey, small and privately owned farms.	High diversification: Cereals, fruits, vegetables, orchards, vineyards, oilseeds, nuts, sugar beets; dairy, pork, sheep, poultry. In Turkey, cotton, olives, figs in addition to products above.	Northwestern part of Balkans entirely rainfed. Albania: 50 percent irrigated. Macedonia, FYR: 15 percent. Turkey: 20 percent. Drought-prone area, hot desiccating winds, intense rain, soil erosion.
Central and Eastern Europe	Current yields low relative to potential. Moldova especially poor and agriculture-based; moderate privatization but highly fragmented private holdings and some remaining inefficient collectives. Privatization also incomplete in Ukraine. In Romania, mix of small family and commercial farms, all privately owned.	Moderate diversification: Wheat, barley, fodder, fruits, vegetables, orchards, potatoes, oilseeds, sugar beets. Livestock, though smaller share than rest of ECA.	Mostly rainfed. Around 10 percent irrigated, except in Romania 30 percent. Moderately drought-prone, Moldova more drought-prone.
Baltics	In Poland, Belarus, and the three Baltic states, farms privately owned.	Little diversification: Barley, rye, wheat, potatoes (especially Belarus). Livestock, pork, poultry. Oilseeds in Poland. Limited fruits, vegetables.	Entirely rainfed, abundant precipitation.
Russia	Farms mostly in Central Russia and Volga region and in N. Caucasus, some in Baltic and in southern Urals and South Siberia. About one-third of agricultural land in private hands, the rest public. Few subsistence farms. Family, joint-stock company farms, and publicly owned farms; low yields from poor management.	Little diversification except in N. Caucasus: Barley, rye, potatoes, fodder in north and west. Spring wheat in north and east, some winter wheat in south. Diverse fruits, vegetables, vineyards in Volga and N. Caucasus. Some rice in N. Caucasus, S. Siberia. Livestock.	Mostly rainfed. Some irrigation in North Caucasus, southernmost part of Urals and Siberia, small amount in Central and Volga. Moderately drought prone in south.
South Caucasus	Most productive arable land now under private ownership, but pasture still communal in places. Small, fragmented holdings. Subsistence and family farms with low productivity.	High diversification: Fruits, vegetables (orchards including apples, pears, cherries, and some citrus), vineyards, cereals, forage, corn, tea. Dairy, sheep.	Armenia, Azerbaijan: 20–30 percent of cropland irrigated. Georgia: 40 percent. Highly drought-prone, but rainfall more abundant in Black Sea coastal area of Georgia.
Kazakhstan	Privatization progressing but incomplete. Small family farms in irrigated south; large farms in the north better-run, private, joint-stock companies growing wheat.	Moderate diversification: Cotton, rice, wheat, fruits, vegetables. Forage, livestock, poultry in south. In north, monoculture of wheat, some oil crops, pasture.	Rainfed pasture. Just 10 percent irrigated. Highly drought-prone, especially in south.
Central Asia	Little privatization, with land ownership and distribution policies distortionary except in the Kyrgyz Republic, where implementing privatization. Subsistence or family farms, inefficient low-productivity collective farms.	High diversification: Cotton, rice, wheat, corn, large number of fruits, vegetables, livestock, poultry, sheep, pasture. Especially reliant on livestock.	Kyrgyz Republic, Turkmenistan, Uzbekistan mostly rainfed pasture. 75–90 percent of region's cropland irrigated. Extremely drought-prone, water-stressed.

Sources: Alam et al. 2005; Csaki, Kray, and Zorya. 2006; FAO 2006b; World Bank 2005, 2008.
Note: In this table, Central and Eastern Europe includes Moldova, Romania, and Ukraine.

Climate Change Is Complicated by Environmental Management Weaknesses

Environmental problems—independent of climate change—have presented substantial challenges to ECA countries, many of which lack management practices needed to protect the natural resource base on which critical economic activities depend (Sutton et al. 2007). Shortcomings are evident in management of soil fertility, water use, pest control, forest health, and illegal logging. Projecting current management practices into an era of accelerating climate change raises concerns not only about social and economic setbacks in farming and forestry, but also about ecosystem stresses, including biodiversity loss and damage to watersheds and rural landscapes.

Failure to address soil erosion is particularly worrisome, since climate change could make today's problems worse through a pattern of alternating droughts and intense rains. Turkey stands out for its progress in managing soil erosion, motivated in part by the wide reporting of the extent of lost output resulting from erosion, which helped to motivate stakeholders. This highlights the importance of monetary estimates to empower champions advocating for change (Sutton et al. 2007).

Institutional and management weaknesses in ECA stem mainly from the complex transition from centrally planned, communist-era governance models. Though the most difficult decade is past, a legacy of distorted specialization and rigid, poorly funded institutions remains. The emphasis on inputs that characterized the region's thinking on agriculture for decades—more fertilizer, more seeds, and more irrigation—have left the sector unprepared to adapt to knowledge-based farming better suited to a world of constrained resources.

Building the capacity to adapt will be crucial for ECA's agricultural knowledge and information systems, which were designed to assist large, public-sector, collective farms in meeting pre-determined production targets for crops and livestock commodities. These systems remain ill-suited for meeting the needs of smaller, private farmers who constitute a large share of the sector today.

Years of over-specialized production have also taken a toll. Under the command economy, collective farms, subnational regions, and even entire countries specialized in an often small number of goods that may or may not have been appropriate to the local natural and human resource endowment. One of the most damaging examples was the concentration of cotton production in Central Asia, which led to overexploitation of water for irrigation, held in place by an institutional framework resistant to diversification.

In the first decade of the region's transition to markets, agriculture, like most sectors, experienced major upheavals, with occasional severe declines in output, and a drying up of government financial support (World Bank 2007). The new private farmers lacked experience in applying modern management and in operating in a market economy. They had little training support from institutions that had either collapsed or remained geared toward the old system. Knowledge gaps combined with a shortage of inputs, equipment, storage facilities, and market structures continue to weaken the farm sector throughout the region (Swinnen and Rozelle 2006).

Yet the agricultural sector is gradually adjusting to policy reforms. Farm economies have begun to recover, with harvests and heads of livestock increasing toward 1990 levels. Private agriculture based on market principles is now predominant. But serious problems persist in the sector's institutional foundations. Environmental laws protecting agriculture, forestry, and biodiversity are weakly or unevenly enforced (Sutton et al. 2007). Systems for research, education, training, and technology transfer suffer from neglect.

Turkey stands out, since it is not emerging from communist-era central planning. Private farms have always dominated agriculture in Turkey, though the small farm size has limited the country's productivity gains. There is diversity in farm production within the country, and agriculture in western Turkey has generally been more progressive and export-oriented than in eastern Turkey. The research, extension, training, information, and technology transfer institutions function relatively well, and cross-ministry cooperation on environmental issues is promising (Sutton et al. 2007).

The capacity to monitor the impact of climate change has largely broken down in Russia as well as in other Eastern European and Central Asian countries, along with services for monitoring baseline weather conditions (see chapter 7). The limited ability to track pests, watch for forest fires, and provide warning of flash floods and other extreme events will increase the risks for farmers and foresters as climate change plays out over time. Because fires pass unchecked across borders, they can spark transboundary political disagreements in addition to causing physical and economic damages. The fires of the summer 2007 in Southeastern Europe offer a sobering example of the human, economic, and political cost of insufficient cooperation and coordinated planning at the national and international level.

Farm Type and Adaptive Capacity

Farmers' ability to adapt to a changing climate depends on the elements of a well-functioning agricultural system:

- timely climate information and weather forecasts, and the skills needed for their interpretation and application

- locally relevant agricultural research in techniques and crop varieties

- training in new technologies and knowledge-based farming practices

- private enterprises, as well as public or cooperative organizations, for inputs such as seeds and machinery, and affordable finance for such inputs

- infrastructure for water storage and efficient irrigation

- physical infrastructure and logistical support for storing, transporting, and distributing farm outputs

- strong links with local, national, and international markets for agricultural goods.

Different types of farms have different degrees of access to these critical elements and, thus, have varying advantages and disadvantages in adapting to the challenges posed by climate change. Although smaller private farms would seem to be the most nimble in responding to changing conditions, larger farms generally have superior climate information and expanded access to credit, and government-owned farms tend to have better access to state sources of information and finance. All else being equal, more diversified operations are better positioned to respond to stresses that might hit one set of crops or one type of activity. Farms already dealing with stressed water supplies will face new hardships in the more uncertain and extreme climate that lies ahead.

Corporate farms in Bulgaria, northern Kazakhstan, Romania, and Russia represent the largest type of farm and have the greatest physical and human capital resources (Csaki, Kray, and Zorya 2006). Next are the cooperative or group farms, generally managed by a few individuals using the pooled land of many smallholders, who may also be hired to provide farm labor. While these farms can exploit economies of scale, their managers typically lack the technological know-how and financing of the corporate farms, making them more vulnerable.

The largest and fastest growing group is the small family farm, which produces for the commercial market but at a small scale. These farms make up the bulk of agricultural income and output in the Balkans, Turkey, the Caucasus, and Central Asia, and remain important in Central and Eastern Europe and Russia. Small family farms will likely continue to serve as the engine of the rural economy in the coming decades, but they may be highly vulnerable to climate change given their size, the farmers' limited technical knowledge, and poor access to public and private information and financial services.

Small farmers in particular will face climate change as yet one more stress compounding many others, including fragmented holdings, marginal land, poor environmental management, ill-defined property rights, increasing demand for standardized and safety-controlled products, declining health and vitality of the rural poor (in ECA, due to aging and outmigration of the young), protectionist food policies abroad, and volatile world food prices (Easterling et al. 2007).

The remaining type of farm is the low-productivity subsistence farm—with aging proprietors supported in part by urban remittances—which have little resilience to shocks. The transition out of agriculture will not be easy for these people since they often have no other options. Safety nets will be needed to assist them.

Potential Climate Change Winners Face Their Own Challenges

Producers and policy makers in northern latitudes have begun to look forward to longer growing seasons and improved farm outputs. However, any complacency would be misplaced, since adaptation investments will be required to take advantage of any potential gains (Parry, Rosenzweig, and Livermore 2005). The potential winners need to be aware of the specific changes projected, how to deal with the uncertainty characterizing projections, and how best to take advantage of the changing climate. Moreover, most countries will have a mix of losing and winning producers, and will require adaptation strategies across sectors and subregions.

New challenges will emerge as producers take advantage of new farming opportunities. Northern areas will see intense competition between forestry and agriculture for land. The relative feasibility of field crops, tree crops, and livestock may further alter land-use patterns. As seen in the case of the Aral Sea, overexploitation of water resources for irrigation, as well as overuse of and resulting runoff from polluting fertilizers, can have devastating consequences on fisheries and other water-dependent activities.

The question of whether ECA's potential winners can realize the benefits of favorable climatic conditions has important implications not just for the countries themselves, but also for world food markets in general. In particular, it is often said that Kazakhstan, Russia, and Ukraine (KRU) have the most unrealized grain production potential, and they could benefit from climate change, at least in their northern regions. But a recent report notes that, since the breakup of the former Soviet Union, these three countries combined have removed 23 million hectares of arable land from production, the largest such withdrawal in recent world history (FAO and EBRD 2008). Almost 90 percent of this land had been used to produce grain.

Bringing large parts of this land back into production could increase world grain supplies and help solve the current global food price crisis. A number of global studies (reviewed in Cline 2007) project a substantial increase in agricultural output for the KRU as a result of climate change (see the caveats in box 5.2). These projected increases contribute to the relatively sanguine attitude of many toward climate change's impact on total world food supplies.

The key question is whether the potential ECA winners will be able to provide the supply response that many expect of them. There are two possibilities for increasing production in the KRU: raise yields on currently cultivated agricultural land, or expand the areas under cultivation. Because the latter would require large investments in land clearing and in infrastructure for production, marketing, and transport, measures to improve productivity of existing farms are more attractive.

The climate, as it is today, is not necessarily the binding constraint on agricultural productivity in ECA—so loosening the constraint by raising average temperatures will not guarantee increased productivity. In other words, productivity depends not only on the climate conditions, but also on technology, investment, support services, and crop management. Analysis has shown that in Central and Eastern Europe and the European parts of the former Soviet Union, the gap between potential yields—under the current climate—and actual yields is significantly higher than any potential gains from climate change. In particular, the current yield gap for the former Soviet areas in Europe (including Ukraine and European Russia) is 4.5 times higher than any potential increase in production from climate change by 2050 (Olesen and Bindi 2002).

While world grain yields have been rising by an average of 1.5 percent per year since 1991, yields in Ukraine and Kazakhstan have fallen, and Russia's have increased only slightly. Yields in all three countries are far lower than those in Western Europe or the United States. The fact that the KRU countries and other ECA countries have not been able to take advantage of this potential for productivity gains suggests fundamental weaknesses in the agricultural sectors of these countries, which does not bode well for their capacity to adapt to and benefit from climate change. Indeed, the immediate challenge is to close the existing productivity gap. Addressing it is a prerequisite for any hopes of riding climate change trends to a new era of prosperity.

Forests show a similar pattern to agriculture. Estimates indicate that the largest share of potential forest stock increases in Eastern Europe would be due to improved management (60 percent to 80 percent) rather than climate change (10 percent to 30 percent) (Easterling et al. 2007). Improved management requires strong forest institutions, which are often lacking in the transition countries.

Adaptation in the Productive Environment

Adapting agriculture and forestry to challenges of climate change—and reaching full potential in today's climate—will demand technologies to monitor and measure conditions in the productive environment, institutions to facilitate knowledge sharing and training, and policies that encourage reform. Managers will need to show resilience and flexibility if they are going to be less vulnerable to changing weather patterns. Many sustainable, appropriately chosen adaptation initiatives would yield measurable benefits regardless of climate factors. Policies and technologies for more efficient distribution and on-farm use of water make economic sense—by lowering costs to government in the form of water subsidies. They also make adaptation sense—by equipping farmers to cope with more variable water availability as well as drought events.

However, adaptation is also a national effort, not limited to individual farmers or foresters. For example, increased water-use efficiency will not be implemented without improvement of existing infrastructure or adoption of new irrigation technologies. And institutional components can be equally important: water-user associations might aid in knowledge sharing, and advisory services can equip farmers with waste-reducing techniques. At the policy level, governments can invest in advisory services and awareness campaigns, while setting water prices to give users incentives to reduce waste and thereby lower government spending on subsidies.

Given the uncertainty about the exact spatial and temporal distribution of climate changes, a cautious approach is to pursue adaptations that would be worthwhile even without climate change. The following examples demonstrate adaptation measures that hold great promise, independent of climate change scenarios:

- *Technology and management* (see annex table 5.1). Conservation tillage for maintaining moisture levels; use of organic matter to protect field surfaces from weather extremes and help preserve moisture; diversification of crops to reduce vulnerability; adoption of drought-, flood-, heat-, and pest-resistant cultivars; modern planting and crop-rotation practices; use of physical barriers to protect plants and soils from erosion and storm damage; integrated pest management, in conjunction with similarly knowledge-based weed control strategies; capacity building for knowledge-based farming; improved grass and legume varieties for livestock; modern fire management techniques for forests.

- *Institutional change* (see annex table 5.2). Support for critical institutions that offers countries win-win opportunities for reducing vulnerability to climate risk and promoting development; key institutions include hydromet centers, advisory services, irrigation directorates, agricultural research services, veterinary institutions, producer associations, water-user associations, agro-processing facilities, and responsive financial institutions.

- *Policy* (see annex table 5.3). Non-distorting pricing for water and commodities; financial incentives to adopt technological innovations; access to modern inputs; reformed farm subsidies; risk insurance; tax incentives for private investments; modern land markets; and social safety nets.

Notes

1. Forestry as a share of GDP is 2.3 percent in Belarus, 0.8 percent in Russia, 1.2 percent in Ukraine, 2.2 percent in Bosnia andHerzegovina, 3.1 percent in Serbia and Montenegro, and about 0.8 percent in Bulgaria, the Former Yugoslav Republic of Macedonia, and Turkey (all figures 2000; Sutton et al. 2007).
2. This point, which was highlighted in the *World Development Report 2008* on agriculture, is well illustrated by the fact that GDP growth originating in the agricultural sector reduces poverty twice as much as growth driven by other sectors (World Bank 2007).
3. Moldova (2000: US$170 million; 2007: US$1 billion), Romania (2000: US$500 million), Bosnia and Herzegovina (2003: US$410 million), Croatia (2003: US$330 million), Albania (1989–1991: US$25 million), the Former Yugoslav Republic of Macedonia (1993: US$10 million) (UNISDR/World Bank 2007; WMO 2007).

Technological Adaptation Practices and Investments for Various Climate, Weather, and Agricultural Phenomena

Technological adaptation practices and investments	Drought	Need for soil-moisture conservation (rain-fed)	Need for water-use efficiency (irrigated)	Land degradation, soil infertility, erosion	Heat stress	Pest and disease control	Excess rain, flooding, storms	Milder winters, longer growing season	Emissions mitigation, carbon sequestration
Land-use management	X	X	X	X	X	X	X	X	X
Mixed farming systems (crops, livestock, and trees)	X	X	X	X	X	X	X	X	X
Conservation tillage	X	X	X	X					X
Nutrient management and use of organic matter	X	X		X					X
Watershed management	X	X		X			X		
Water-harvesting techniques, storage, reduction of runoff	X	X	X	X	X				
Drainage systems				X		X	X		
Rehabilitation and modernization of irrigation infrastructure, canals	X		X		X		X		
Introduction of new irrigation facilities	X		X		X		X		
Use of marginal water	X		X		X				
Dams for water storage, flood control	X		X	X	X		X		
Supplemental irrigation	X		X		X				
Irrigation only at critical stages of crop growth	X		X	X	X				
Sprinkler irrigation	X		X	X	X				
Drip irrigation	X		X	X	X				
Furrow and flat-bed irrigation	X		X		X				
Crop diversification	X	X	X	X	X	X	X	X	
Use of water-efficient crops, varieties	X	X	X		X			X	
Use of heat- and drought-resistant crops (varieties and hybrids)	X	X	X		X			X	
Switch to crops, varieties appropriate to temperature and precipitation	X		X	X	X	X	X	X	
Crop rotation (sequencing)	X	X				X			
Switch from field to tree crops (agro-forestry)	X	X	X	X	X	X	X		X
Timing of operations (planting, inputs, irrigation, harvest)	X	X	X		X	X	X	X	
Strip cropping, contour bunding and farming	X	X		X			X		
Vegetative barriers, snow fences, windbreaks	X	X	X	X			X		X
Rangeland rehabilitation and management	X	X		X	X	X	X	X	X
Pasture management (for example, rotational grazing) and improvement	X	X		X	X	X	X	X	X
Supplemental feed	X				X				
Fodder banks	X				X				
Watering points	X			X	X			X	
Livestock management (including animal breed choice)				X	X	X	X	X	
Fire management for forest and brush fires	X			X	X	X			X
Response farming (using seasonal forecasts)	X	X	X	X	X	X	X	X	
Integrated pest management	X					X		X	

Sources: Sutton, Block, and Srivastava 2008; Padgham 2008.

ANNEX TABLE 5.2
Institutions critical for adaptation

Institution	Importance for adaptation	Status in ECA
National and local governments		
Hydromet and forecasting centers	After essential information for planning; understanding changing climate; and providing farmers with long-term, seasonal, and daily weather forecasting for knowledge-based response farming.	The former Soviet Union was served well, but service has since crumbled. Centers are improving in European Russia but are still unsatisfactory in Central Asia and the rest of ECA. Centers have poor capacity for local monitoring, local data interpretation, and forecasting.
Advisory services (including agricultural extension)	*Agronomic information.* Interpret hydromet output for practical advice to farmers; convey information on trends of climate change and risk; recommend and train in new and off-the-shelf technologies and in new or different locally adapted crops and varieties; demonstrate new farming practices. *Financial advice.* Provide information on sources of finance for adaptive investments. *Market information.* Provide information on market prices and channels of distribution for crops and livestock (key to ensure that services reach small- and medium-size family farms).	Generally, both public and private sector advisory services are in poor condition. Challenge is to reach small farmers. They lack of capacity for interpretation of climate forecasts, interpretation of probabilistic climate data, and, thus, communication of probabilistic and not deterministic forecasts. In Turkey, advisory services are better developed but lack capacity to effectively advise farmers in an environment of increased challenges.
Irrigation directorates	Maintain, rehabilitate, expand, and replace old and new irrigation facilities, which will be more important in water-stressed areas. Intermediary between managers of water resources and farm users.	
Forestry departments and agencies	Maintain health of forests, and respond to pests and risks of fire. Observe changes in forest ecosystems in response to changing climate. Participate in planning related to forest–agriculture land trade-offs.	In much of ECA, these institutions are often among the best functioning of those that will be relevant for climate adaptation.
Agricultural research institutes	Bring knowledge of locally relevant needs to research networks at local and international levels, develop varieties and technologies suitable for changing climate and local endowments.	After the disintegration of the Soviet Union, research systems collapsed and are not effective in meeting the current demands. In Turkey, the situation is better.
Agricultural education at vocational schools and technical colleges	Provide important conduit for information about implications of climate change for farmers and managers, including adaptation measures and technologies and guidance on how and when to implement them (key in move toward more knowledge-based rather than input-based farming).	
Quality control, phytosanitary, and veterinary services	Provide standards information and enforcement consistent with national and international regulation, monitor and control livestock health, and provide timely information on disease risks.	Services are strong in some countries, but not up to challenge of global food market in others.

Institution	Importance for adaptation	Status in ECA
Civil society		
Producer associations and farmer organizations	Share information about outcomes and challenges of adaptation, serve as locus for absorbing new information from and communicating farmer concerns to government bodies and private enterprises, allow shared investment in new machinery by small farmers.	Producer associations and farmer organizations are starting to grow, and their effectiveness varies across countries. There is potential for further expansion to more areas and for deepening of activities.
Water-user associations	Encourage more sustainable water use.	They are relatively recent institutions, not fully developed and just beginning to function.
Nongovernmental organizations	Provide information, funding, and institutional support at small scale for pilot adaptation efforts by farmers, offer microcredit to enable adoption, share knowledge of local experiences, advocate farmers' concerns.	Their moderate presence is increasing in ECA countries. They face the usual challenges, such as interventions not sustained after projects end, struggle to reach the neediest, lack of coordination with other organizations.
Private enterprises		
Private and public seed companies and nurseries	Ensure production and availability of seeds and seedlings of appropriate varieties (e.g., with improved drought- and pest-resistance), to take advantage of agricultural research and development and facilitate adoption.	In Europe, they are available but currently inadequate. They have a limited presence, efficacy in Caucasus, Central Asia. They are good in Turkey.
Grain storage and drying facilities	Will be needed in currently unserved, newly cultivated areas, and areas with intense rainfall or heat, which cause rot and spoilage.	They are not present or are inadequate in areas that will need them as cropping and livestock zones shift, and as rainfall increases during cereal harvesting time in the Baltics, Central and Eastern Europe, Russia, northern Kazakhstan.
Agro-processing facilities	Offer processing of livestock products in expanded pasture areas, processing of horticulture crops in new areas.	They are not present or are inadequate in areas that will need them as cropping and livestock zones shift northward.
Marketing enterprises	Exploit economies of scale by buying produce of family farms and selling at market, which mitigates risk to farmers of adopting unfamiliar crops or varieties with uncertain local demand.	They are variable and with scope for improvement. They are generally stronger in Turkey and Europe than in the Caucasus, Central Asia.
Financial services	*Banks.* Provide necessary finance for implementation of adaptations. *Microloans.* Reach out to small farmers with limited access to formal banks. *Agricultural insurance.* Mitigate risks of crop failure from unpredictable weather, unproven adaptations, market uncertainties.	Small farmers have poor access to banks. They have limited presence, effectiveness of microcredit organizations. Weather-indexed insurance does not exist in most ECA countries.

ANNEX TABLE 5.3
Policies Critical for Adaptation

Policy	Importance for climate change adaptation and implementation challenges
Non-distortionary water pricing	Reduce subsidies to increase incentives for more productive allocation, management, and use of water resources. Reduction is difficult because removing subsidies often meets political resistance.
Non-distortionary commodity market policies	Reduce distortions in markets for cereals and oilseeds, including setting price caps and taxing or otherwise restricting exports. Letting prices pass through will increase incentives for producers to invest and expand production of these crops over time. Export restrictions become contagious, significantly reducing agricultural trade and the ability of world food markets to respond to climate change. Manage state grain reserves transparently and effectively in order to ensure supply during short-term shocks, rather than to keep prices low.
Financial incentives for adoption of technological adaptations	Provide tax incentives for measures such as farmers' purchases of machinery required for conservation tillage, or planting of drought-resistant seedlings. Provide financing and coordination for hiring of machines and labor for reforestation projects.
Access to modern inputs	Remove restrictions on imports of modern seeds and seedlings to allow farmers access to modern varieties (for example, with increased drought resistance or longer maturation).
Invest in support institutions (identified in annex table 5.2)	Provide funding in many countries, where these institutions have been underfunded for a long time. Both priority placed on investing more and availability of public finances for doing so vary across countries.
Reform farm subsidies	Avoid trying to "pick winners" for example, subsidies for cereals rather than the fruits and vegetables that may become more appropriate due to warming. Subsidies targeted at production of specific crops may be counterproductive as comparative advantages change. Recurrent production subsidies also divert resources from potential investments in public services and farm investment (not production) subsidies.
Promote private investments	Promote investments by the private sector in new technologies by providing tax incentives, matching grants, and technical assistance, not only for primary production, but also for inputs, processing, logistics, warehousing, and other related activities.
Insurance	Explore opportunities for developing system of weather index insurance (as opposed to traditional multi-peril crop insurance). For smaller countries especially, spread risk across countries.
Improve land markets	Ensure land tenure security, improve land registration and cadastre systems, and reduce market transaction costs. This will help to increase the flexibility of farmers, reduce fragmentation, increase access to finance, and encourage investment.
Calculate economic costs and benefits	Calculate the economic costs and benefits of policy changes and investments decisions as rigorously as possible to ensure the most efficient and effective use of public resources. This will often require capacity building.
Encourage livelihood diversification	Provide training and financial support to encourage the development of non-farm rural employment or skills for urban employment. In some areas, and for some rural residents, agriculture and forestry may become unviable.
Strengthen social safety nets	Provide targeted income support for poor and vulnerable segments of the population that may have difficulty affording food, who may live in areas where agriculture becomes unviable, or who may not be able to easily change livelihoods (especially the elderly or sick).

CHAPT

The Built Environment: Cities, Water Systems, Energy, and Transport

JoAnn Carmin, Tim Carrington, Jane Ebinger, Barbara Evans, Franz Gerner, Bjorn Hamso, Antonio Lim, Ziad Nakat, Ana Plecas, Michael Webster, and Yan F. Zhang

The housing and infrastructure in Eastern Europe and Central Asia (ECA) is acutely vulnerable to physical changes from climate variability and extremes. Floods are an obvious threat in many cities. Storm surges in the Black Sea and elsewhere are affecting coastal infrastructure. Projected warming trends and changes in precipitation patterns have the potential to impact the entire energy chain—from production, through transmission and distribution, to end use. With the increasing likelihood of many more extreme events—floods and droughts—water quality could be profoundly affected.

This vulnerability is driven mainly by the poor condition of the infrastructure. Relatively fewer stresses are needed to overwhelm old, badly maintained or constructed installations. Housing provides one example. From the mid-1950s through the late 1980s, state enterprises built multi-story, multi-family housing blocks from

This chapter was drafted by Tim Carrington, based on four background papers prepared for this book, "Europe and Central Asia Region: How Resilient is the Energy Sector to Climate Change?" by Jane Ebinger, Franz Gerner, Bjorn Hamso, Antonio Lim, and Ana Plecas; "Adapting to Climate Change in Europe and Central Asia: Background Paper on Water Supply and Sanitation" by Barbara Evans and Michael Webster; "Achieving Urban Climate Adaptation in Europe and Central Asia" by JoAnn Carmin and Yan F. Zhang; and "Climate Change Adaptation in the Transport Sector" by Ziad Nakat.

prefabricated concrete panels, most of them designed for a life of about 30 years. In Poland, for example, there are more than 5 million Soviet-era flats, many in desperate need of refurbishment. When additional stresses of higher winds, more intense precipitation, summer heat waves, or melting permafrost in some regions are added, some of the buildings could become even less livable. Transport systems, energy infrastructure, and water utilities are similarly vulnerable.

While the most significant impacts of climate change are perhaps decades away, some vulnerabilities are already evident. A flood in Baia Mare, Romania, in 2000 brought cyanide-laced waste from a gold mining operation into the Tisza and Danube rivers, tainting the drinking water of 2 million people downriver (Carmin and Zhang 2008). The mix of extreme weather and past environmental mismanagement turned a flood into a major threat to public health. Storm surges in the Black Sea are affecting coastal settlements, and more severe conditions may damage the 23 ports along the sea coast. The more extreme heat conditions of Central Asian summers have exacerbated problems of poor road maintenance and low design standards. Warmer temperatures and resulting ground settlement in permafrost areas of the Russian Federation have destabilized a number of structures, including residential buildings, a power station, and an airport runway in Yakutsk, Russia (Ebinger et al. 2008).

How well the cities, buildings, and infrastructure of the ECA region can cope with climate change will depend on whether governments improve current management practices and address quality deficits that leave so many structures vulnerable. Barring runaway catastrophes, climatic changes are likely to be manageable if utilities and structures are well run and maintained. But it takes far smaller shocks to overwhelm overstretched utilities, decrepit housing, and poorly maintained infrastructures. Policy makers must identify the most vulnerable structures and accelerate, retrofit, and upgrade programs to improve their energy efficiency and livability while increasing their resilience to the effects of warmer and more extreme weather patterns.

Following are analyses of the impacts of projected climate change on urban structures, energy service provision, water systems, and transport infrastructure. Each shows some of the ways that a warmer, wetter, and more extreme climate may affect existing structures and systems and suggests a framework and practical steps to lessen the risks. Proposed actions would support sturdier, better maintained structures and assist governments to protect buildings, roads, ports, energy systems, and waterworks from the damaging effects of climate change.

Urban Challenges: Making Cities Livable and Viable in a Warmer Century[1]

About two-thirds of ECA's population lives in cities (though fewer in Central Asia and the South Caucasus; see table 5.1), and many cities are beginning to experience the effects of climate change. Some are encountering water shortages; others are facing increased or more variable precipitation, rising temperatures, or more intense extreme weather events. Over time, continued shifts in weather patterns could damage some buildings and make others uninhabitable, stress infrastructure, threaten urban plant and animal life, and increase illness and deaths among vulnerable populations (box 6.1).

Despite the potential risks for cities and their residents, few municipalities in the region have integrated climate adaptation into their planning. To increase the resilience of cities to projected changes, ensure their livability, and maintain the provision of basic services in the long term, local governments need to begin planning today. Plans will have to address issues such as projected higher temperatures in the summer months, associated increases in pollution and heat outdoors, and altered indoor air quality and temperature in many buildings. While this may be less problematic in the far north, the increased incidence of heat waves across southern and central Europe will require that buildings have improved ventilation and

---

BOX 6.1

Roma, Already Marginalized, Are Particularly Vulnerable

Across the world, marginalized communities remain the most vulnerable in times of natural disaster. In the former East Bloc, the Roma—dispersed across the region—face continual stresses. Not only are many Roma neighborhoods overcrowded, but a study conducted in 2000 in Bulgaria, Hungary, and Romania found that most homes also do not have hot running water or central heat and showed an overall state of disrepair (Revenga, Ringold, and Tracy 2002). When floods hit the Slovakian town of Jarovnice in 1998, approximately 140 Roma homes were affected and 45 Roma died, compared to 25 non-Roma homes and 2 non-Roma deaths. Similarly, when the floods of 1997 hit the Czech city of Ostrava, white, non-Roma residents were offered opportunities to resettle in flats outside of the flood area, while Roma families were offered small workers' cabins or sent back to their flooded homes, even though the homes were in an area deemed unfit for habitation.

Sources: Adapted from MRG 2008; Bukovska 2002.

cooling, not only for those individuals most vulnerable to health threats from the heat—the elderly, infants, and the disabled—but for the general population as well. In southern cities, projected reductions in precipitation and higher temperatures could also lead to groundwater depletion. In addition to raising concerns about water shortages for urban dwellers, reduced moisture in soils can affect the foundations of buildings.

Another issue that cities face, and which climate change can aggravate, is the urban "heat island" effect. Most urban areas are built with surfaces that absorb the heat interrupted by parks and green spaces populated with plants that are suited to historic climate patterns. As temperatures increase, some plants may have difficulty surviving the new climate. When the amount of non-reflective surfaces in cities combines with the heat generated through rising energy use, cities can become significantly warmer than surrounding areas, raising concerns about heat stress and unmanageable surges in energy demand for cooling.

Coastal cities face additional concerns of infrastructure vulnerability (see chapter 4). Sea-level rise, and the associated increase in storm surges, will accelerate coastal erosion, increase the incidence of flooding, and lead to saltwater intrusion into groundwater aquifers in cities, particularly those along the Baltic and Adriatic seas. Turkey, for example, is highly vulnerable since it is bordered by four seas (Aegan, Black, Marmara, and Mediterranean). A 1–meter rise in sea level would affect approximately 30 percent of the nation's total population living in urban areas in proximity to the coastline. Sea-level rise has the potential to affect not only natural systems and housing and infrastructure, but also tourism and enterprise (Karaca and Nicholls 2008).

Many northern cities situated along major waterways face the prospect of greater precipitation, leading to river swell and stress on existing dams. As precipitation increases and soils become waterlogged, existing stormwater drainage systems, as well as sewage treatment plants and sewer lines, may be overwhelmed. Sewers that carry both storm water and sewage are common in many cities throughout the region. During the Prague floods of 2002, these systems were overwhelmed, and many sewage treatment plants had to halt operations. Flood waters can transfer contaminants from abandoned industrial sites and operational facilities to populated areas. Along with the other types of wastes that will wash up onto the shores, these conditions can pose threats to human health.

Large, prefabricated, and poorly maintained Soviet-era buildings, a dominant feature of so many cities in the region, are vulnerable to

TABLE 6.1
Projected Refurbishment Needs Relative to Support Programs

Aspect	Latvia	Poland	Lithuania	Estonia	Eastern Germany
Number of flats in panel buildings, built 1950–90	416,460	5,200,600	790,000	406,570	2,150,000
Assumed average refurbishment requirement per flat (US$ millions)	11,600	11,600	11,600	11,600	29,000
Overall refurbishment requirement (US$ millions)	4,831	60,327	9,164	4,717	62,350
Investments achieved with support programs (US$ millions)	4	363	29	44	43,500
Refurbishment covered to date by support programs (%)	0.10	0.60	0.32	0.92	69.77

Source: BEEN 2007.
Note: Original estimates in euros were converted to U.S. dollar amounts at an exchange rate of US$1.47 to €1.

projected changes. The housing stock is often undermaintained, energy inefficient, leaky, and a visible weakness in the urban fabric. Table 6.1 shows the extent of prefabricated panel flats and the projected costs of refurbishing the buildings.

Formulating plans and mobilizing resources for retrofitting work is a priority across the region. Ideally, retrofits should draw on sustainable technologies to provide for healthier interior conditions and sturdier resistance to extreme weather, while reducing carbon emissions through energy-efficient systems, thereby helping to reduce costs for consumers, spikes in energy demands for cooling, and the emissions driving the overall climate change problem.

Retrofitting on a large scale is costly, but the technologies and solutions are straightforward. Much of the retrofitting taking place in ECA and elsewhere focuses on energy-saving measures. These include thermal insulation, replacement windows, and modernization of central heating systems. In addition to these measures, green roofing is being tested as a further means for improving the quality of the living spaces as well as for managing fluctuations in precipitation (box 6.2).

In recent years, the region has seen increases in urban sprawl, altering the profile of urban vulnerability. As cities move to develop adaptation plans, city managers and planners could promote new, compact, and sustainable construction and site planning and zoning policies that reflect climate change risks. For example, government

BOX 6.2

Green Roofs to Manage Stormwater and Heat Waves

A green roof is a roof partially or completely covered with vegetation and soil, planted over a waterproofing membrane. It may include additional layers such as a root barrier and drainage and irrigation systems.

Green roofs are increasingly popular for two reasons. First, they help stormwater runoff management: they retain up to 75 percent of rainwater, gradually releasing it back into the atmosphere via condensation and transpiration, while retaining pollutants in the soil. They also help combat the urban heat island effect. Traditional building materials soak up the sun's radiation and re-emit it as heat, making cities much hotter than surrounding areas. Green roofs can cool the surrounding air by as much as 3°C to 11°C at the same time that they reduce the need for air conditioning inside the building.

Green roofs have been around for thousands of years (from the sod roofs of rural cabins to the hanging gardens of Babylon) but are now making a major comeback. Germany pioneered their modern incarnations in the 1970s, when existing sewage systems were unable to cope with heavy rains. Now, many local authorities in Austria, Germany, and Switzerland require that new buildings include them. Green roofs are becoming more common across Eastern European countries—a well-known example is that of the Warsaw library.

Sources: Taber 2008; Green Roofs for Healthy Cities, http://www.greenroofs.org; Gill et al. 2007.

policies can enhance the hydrological environment's natural ability to adapt by limiting development in areas affected by high precipitation, flooding, or other weather-related events, or by preserving green spaces and waterways. Site planning must extend to consider industrial areas, mining operations, and brownfield sites to address the risks that these areas pose to people and settlements when floods occur. In addition, new building codes and energy conservation ordinances should be aligned with principles of green design.

Operating from a planning paradigm that incorporates climate change will require new processes and new capacities. Municipal governments and government agencies must have the capacity to plan for and implement adaptation measures. *Capacity* in this case refers to technology, expertise, financial resources, staffing, and interagency coordination. Given the nature of climate change, there must be strong ties to the scientific community so that timely information is received. There should also be mechanisms that retrieve input about changes from local communities so that officials can respond.

Local communities must be part of the decision-making process (chapter 1), and lessons should be drawn from cities already engaged in adaptation planning (Prasad et al. 2009). In addition, future research can make a significant contribution. Questions that could be explored include the following:

- Which cities are most vulnerable to the impact of climate change?

- Where are the vulnerable populations located, and what steps can be taken to reduce their risks?

- What are the drivers for municipalities to initiate climate adaptation planning and action?

- What municipal adaptation planning efforts have been most successful, and which problems have surfaced frequently?

Supplying Water: Essential for Human Activity and Facing Multiple Pressures[2]

Extreme precipitation, drought, and heat waves can all have negative impacts on water quality. For example, floods often bring about wastewater overflows and contaminated runoff from farms and factories. Increased sediment loading may occur in areas already stressed from deforestation, resulting in increased water treatment costs. Where drier weather and drought cause a decline in flows in lakes and streams, there will be increased concentrations of pollutants and changed biological properties in the water sources upon which communities rely. Hotter days bring increased surface evaporation, leading to greater salinization. Sea surges lead to saltwater intrusions in coastal aquifers.

While climate change promises a mélange of effects—some positive, such as longer growing seasons in northern regions—the fallout for water systems is overwhelmingly negative. Water professionals are confronted with an expanded set of possibilities and extremes and face more complex choices. Where water is less available, communities will have to change their water-consumption patterns or bring in water from farther away. Hydropower output could be affected by varied or lower inflows in some regions, straining energy supplies. Stormwater drains may prove inadequate.

In general, water structures such as pipelines, reservoirs, and dikes have been designed on the basis of historic climate trends—but new climate patterns may call for structural shifts. Simple calculations of supply and demand raise other concerns. Population growth combined with increased agricultural and industrial demands may coincide with diminishing water resources, particularly in Central Asia. In other parts of ECA, heavily populated coastal areas already face an array of pollution and groundwater problems that will only worsen. Sea surges will instigate more mixing of saltwater in aquifers and less available fresh water. Throughout ECA, there is the continued risk that sewage and inorganic materials will mix with water supplies.

Most water utilities in ECA face additional challenges that hamper their capacity to adapt. Overstretched and underfunded, water and sanitation utilities show relatively poor performance, and most cannot cover their costs. The results include shortcomings in service delivery, quality, and capacity, some of which are described as follows.

• *Lower-than-expected coverage—particularly in rural areas.* Although the ECA region has nominally high access to improved water sources and sanitation, 27 million people still lack access to improved water supply (WHO and UNICEF 2006). In addition, quality and reliability are often poor. Even in capital cities, possibly even less than 65 percent of connected households enjoy a 24-hour supply, and performance is typically worse in smaller towns. According to a recent study (OECD 2005), the trends in the water supply and sanitation sector suggest a continued further deterioration of water services, even without climate change.

• *Highly inefficient systems with low revenues and high investment needs.* Non-revenue water rates are high (physical losses alone are in excess of 40 percent in eight countries of the region) as are labor costs (most utilities report three to five staff members per thousand connections, which can be compared with the United Kingdom average of 0.3–1.0 staff members per thousand). Cost recovery is often low, with water utility revenues across the region estimated to cover only around 60 percent of operational costs—for example, 61 percent in Russia and 64 percent in Ukraine (OECD 2005). This is due to a combination of unwillingness to raise tariffs and expensive Soviet-era designs. The low revenue base translates into a cycle of underinvestment, poor maintenance, deterioration of infrastructure, and rising costs. Resources for rehabilitation and major investment are scarce, and the poor revenue record makes borrowing difficult. An estimated US$15 to US$34 per capita per year of additional finance is needed simply to maintain infrastructure at its current levels (OECD 2005).

• *Transition from centralized economies to municipal government.* Most countries in the region have undergone a rapid and almost complete decentralization to the municipal level, placing severe strains on local government capacity and finance. The resulting underinvestment may have had a knock-on impact on technical skills and capacity within utilities.

In general, water utility planning in ECA is only weakly linked to the overall management requirements for water resources as a whole (World Bank 2003)—although there have been notable successes in the Baltic Sea states and slow progress is being made in the Aral Sea basin. Changes are clearly needed to create stronger incentives in the water supply sector through stronger linkages to water resource management and greater efforts to stimulate capital flows to cash-starved utilities. Pilot programs in managing water markets will be useful, in addition to further research to identify the most vulnerable systems.

To address the above shortcomings and improve climate resilience, governments could explore practical steps to improve efficiency in the short term and the long term and lower sensitivity to climate-related disruptions. Possible priorities include the following:

- *Improve demand-side management.* There is considerable potential to reduce water demand; in most ECA countries consumption levels remain high by international standards. Improved metering and tariff-setting are critical. In parallel, water supply infrastructure could be rehabilitated to significantly reduce losses. Cutting water consumption through a variety of conservation measures and efficiency improvements would not only reduce vulnerabilities in the water sector, but also save significant amounts of energy.

- *Improve water storage.* Constructing new dams and reservoirs to increase storage would help those countries facing probable droughts and exhaustion of water supplies. A lower-cost option is to improve the management of existing reservoirs and dams.

- *Improve flood protection and drainage systems.* Investment in flood protection will be important for dams, treatment plants, and distribution systems, while improved storm drainage could limit flood damage and protect groundwater supplies.

- *Explore the benefits of desalination facilities.* Desalination has long been a costly strategy for expanding water supply. However, with high costs for alternative supplies, this option may become more attractive in light of changing climate scenarios, particularly if it is reliant on renewable energy such as solar power.

The process of evaluating these and other possible investments demands a capacity to make sound economic judgments about costs, risks, and trade-offs. Climate change calls for new and sophisticated planning skills, which many of the region's utilities lack.

Finally, the significant variation in exposure and sensitivity across the region implies a need for locally determined adaptation plans. While planning models can be similar, each locality must be able to analyze specific risks and to fashion programs that address the most urgent threats.

Energy-Sector Vulnerabilities: New Pressure to Overcome a Legacy of Inefficiency[3]

The supply, transmission, and distribution of energy will be affected by the changing climate, particularly as the region experiences more climate variability and increasing episodes of extreme weather, such as droughts and flash floods. And although this book focuses on adapting to climate change, the region is a contributor to global warming and synergies exist among the goals of reducing the region's carbon footprint and helping the energy sector adapt to new and more challenging climate conditions (box 6.3).

Rising temperatures across the region will lead to changes in the level and timing of peak demand, resulting in a flattening of the electricity consumption profile across the year as demand for cooling energy rises and heat energy declines. While ECA-specific projections are unavailable, European data are indicative: heating demand is projected to decline by 2 to 3 weeks per year, and cooling demand is projected to rise by between 2 to 3 weeks (in coastal areas) and 5 weeks (in inland areas) by 2050. In the Mediterranean, these changes should result in a decrease in heat energy demand of up to 10 percent, but an increase in cooling demand of nearly 30 percent (Parry et al. 2007).

Other potential climate-related concerns for the energy sector in particular subregions include the following:

- *Lower heating costs, higher cooling costs.* The trade-off accompanying warmer winters, with lower demand for heating, is a costly increase in demand for cooling. The Baltic countries, along with Poland and Belarus, will likely see lower need for natural gas and electrical power imports. But more days of extreme heat—above 35°C or 40°C—could place new burdens on power systems in southern and eastern parts of ECA, particularly for cities, which will experience increased temperatures due to heat island effects. Electricity systems—some already stretched, such as those in Central Asia, Southeastern Europe, and Turkey—may struggle to meet heavier demands for air conditioning, particularly if they rely on hydro-

BOX 6.3

ECA's Energy Sector in Need of Investments and Improved Management

The ECA region, accounting for 5 percent of the world's gross domestic product (GDP) but 10 percent of its energy demand, is the world's most energy-inefficient region in terms of both consumption and production of energy (see figure). Sector assets employ old and outdated technologies, many running beyond design life; the average age of power generation facilities is 35–40 years with nearly 80 percent installed prior to 1980. Poor maintenance throughout the 1990s has left systems more inefficient, unreliable, and polluting.

ECA has the world's highest carbon intensity
Total primary energy supply in ktoe (kiloton oil equivalent) per GDP

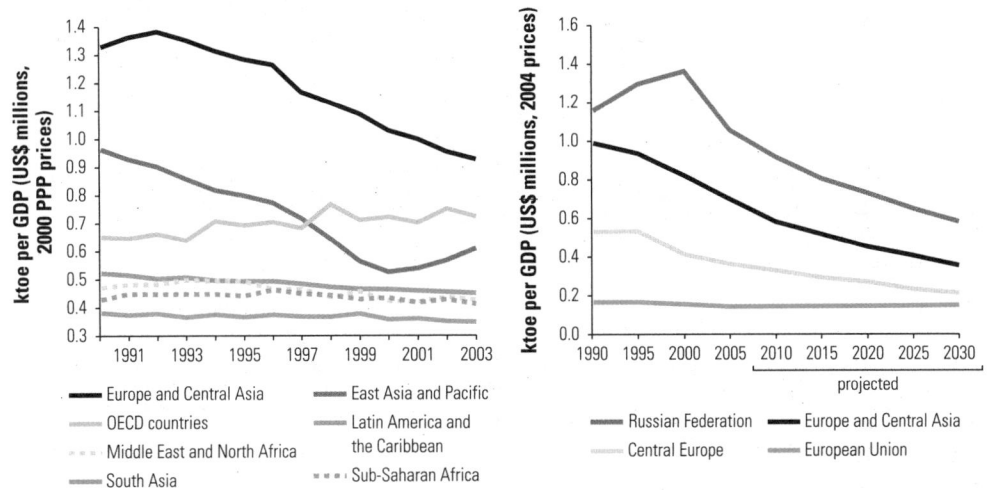

Source: Ebinger et al. 2008, with data from WDI, IEA, and ECA Energy Flagship Model.
Note: PPP = purchasing power parity.

Demand is expected to rise in the period to 2030—electricity consumption grows at an average annual rate of 3.7 percent—and fossil fuels are expected to remain the dominant source of energy. Future gas and electricity shortages are possible in several subregions (Central Europe, Southeastern Europe and Turkey, and Russia) threatening rapid growth. Together with rising gas prices and concern about reliance on Russia for fuel, the region is tending toward a growth pattern based on more polluting (but locally available) coal and resistance to shutting down aging nuclear reactors.

By 2030, coal-fired and nuclear generation are both projected to increase to 35 percent and 20 percent, respectively, of total generation, while hydropower and gas-fired generation will decline to 12 percent and 29 percent, respectively. Expectations are that about half of today's infrastructure will be rehabilitated by 2030, while 40 percent is retired and around 726 gigawatts of new generation capacity is built, mostly thermal (72 percent). Overall, investment costs are estimated at US$1.2 trillion. The renewal of sector assets in the period to 2030 provides a window of opportunity to curtail the carbon footprint and increase the resilience of the sector to climate change.

Source: Ebinger et al. 2008.

power, which could become scarcer just as it is needed most, during periods of heat, accelerated evaporation, and drought.

- *Altered contribution from hydropower.* Hydropower in Southeastern Europe (including Turkey) and Central Asia will see changes in the timing and volume of flow to storage systems. Runoff will significantly decline (in some parts up to 25 percent), but in the near term this decline may be balanced by increased water availability from more rapid glacial melt in the Alps, the Caucasus, and the mountains of Central Asia. The melting will initially increase stream flow, which is then expected to decline over time by up to 50 percent in some areas as snowpack and glacier formation fail to keep pace with rates of melting. Hydropower potential around the Mediterranean is projected to decline by 20 percent to 50 percent while increasing in Eastern Europe by 15 percent to 30 percent and remaining stable in Central Europe (Alcamo et al. 2007).

  Changing conditions will affect generation efficiency (sedimentation), reservoir management (storage and use, mudflows, and lake outbursts), and seasonal water availability. There may be increased competition with other sectors or neighboring countries for scarce water supplies. At stake may be water-export arrangements between the Kyrgyz Republic and Tajikistan—both comparatively rich in water resources—and drier Kazakhstan, Turkmenistan, and Uzbekistan. However, northern parts of Europe and parts of Russia will see increased hydropower capacity.

- *Pressures on thermal and nuclear power.* The operation of thermal and nuclear power facilities will be challenged by availability and temperature of water because of their dependence on significant volumes of water for cooling. Lower water levels in lakes and rivers, reduced runoff, accelerated evaporation, and warmer water could limit availability of water for cooling or cause restrictions on cooling water intake or discharge, constraining facilities' generation capacity. This stress could translate into interrupted and more expensive electricity generation. Impacts are likely to be less significant than for hydropower, but will still require new operational management strategies and considerations in design.

- *Extreme weather effects on network management.* Extreme weather stretches the abilities of power transmission networks to function, reducing efficiency or impacting structural integrity, particularly for older and poorly maintained facilities. Transmission capacity, already constrained in parts of Russia, Southeastern Europe, Central Asia, and the Caucasus, may be hampered by load management issues, especially during summer peak demand. Efficiency can decline with rising temperatures because of issues such as line sag and extreme events that affect line integrity, including heavy snowfall, precipitation, windstorms, and icing.

- *Mixed impacts for extractive activities in Arctic and Siberian Russia.* Rising temperatures in Arctic and Siberian Russia, which could open up major economic opportunities (such as offshore oil exploration), will have negative impacts in zones of discontinuous permafrost.

  Oil and gas extraction and mining in permafrost areas will have to adjust to changes, including new challenges from thawing and shifting ground (see map 4.2). Freeze–thaw processes already have a negative effect on the structural integrity of buildings, key infrastructure (access routes, power plants, and mines), and pipelines, leading to failing pilings, heaving structures, and eroding shorelines and riverbanks. For example, collapsing ground in Yakutsk in western Siberia has already damaged several large residential buildings, a power station, and a runway at the Yakutsk airport. Thawing and ground settling are impeding railways and roads used in energy transport, reducing the number of access days for transit routes and operations sites.

  In offshore areas, reduced sea ice will lengthen the navigation season, allowing exploration and exploitation of as yet untapped mineral resources and reduce costs for industries that rely on shipping for transit. However, broken-free sea ice and increased storm surges may endanger shipping, accelerate coastal erosion, and increase the risk of pollution.

- *Vulnerability to floods.* More frequent flooding from rivers in the interior or from sea surges threatens all types of structures, including energy infrastructure. In Romania in 2005, six consecutive waves of flooding led to widespread power cuts. Structures near coastlines—such as a Russian oil storage facility on the barrier island of Varandei in the Pechora Sea—are already under threat because of changing sea levels.

- *Opportunities for renewable energy.* Projected higher wind speeds bring new opportunities for wind-power generation, both offshore and inland. In addition, more solar power may become more feasible for Mediterranean areas. But wind and solar power are also sensitive to climate phenomena—such as more variable wind patterns and increased cloud cover during warm months.

  From an adaptation perspective, the key question for regulators and industry alike is how much to invest in adaptation today given the uncertainties in climate forecasting and the impact, in the coming decades, of greenhouse gases already in the atmosphere.

A growing number of specialists now support a risk-based and flexible approach that focuses on no-regrets and win-win adaptation solutions, combining infrastructure investment with operational management solutions and further monitoring and research (see chapter 1).

Despite many unknowns, it is certain that ECA's energy sector will be affected by climate change, although the nature and degree of impacts will vary across the region. On the positive side, the energy sector is accustomed to working in harsh environments, adapting—at a cost—to the realities that present themselves. The oil and gas industry has a long history of working in harsh environmental conditions and seeking innovative technical solutions to operational challenges. The power sector has vast experience in day-to-day grid management operations based on short-term climate forecasting. Most adaptation measures are already known, and the resilience and resourcefulness of the sector will be important assets. However, financing could present a constraint. Future strategies will have to include and engage a broad range of stakeholders, who will be affected both by climate change and by the various schemes to adjust to it. Some options to address management and structural issues include the following:

- *Transfer best practices.* Best-practice technical solutions developed for the energy sector in other parts of the world could be transferred to ECA. For example, North American experience offers potential solutions for issues facing Russian Arctic and Siberian permafrost zones today.   .

- *Look at demand side management.* Energy saving and demand-side management measures provide a cost-effective, win-win solution for mitigation and adaptation concerns in a context of rising demand and supply constraints. Water resource and flood management techniques are well known and will be important for those regions suffering drought conditions; meanwhile, regional cooperation, integration, and trade (for energy and water) can offer potential solutions as well.

- *Optimize the design for new or retrofitted investments.* The anticipated large investment in ECA's energy infrastructure in the coming decades provides a window of opportunity for smart, climate-resilient design. Targeted refurbishing can help solidify weaker elements of the energy infrastructure assets that have a typical lifespan of 30 years to 50 years. Meanwhile, investment in design standards to reflect projected changes can increase the resilience of new infrastructure. For example, where permafrost is melting, deeper pilings can be used, and buildings can be raised slightly above the ground and thickly insulated. Lighter-weight building materials can be employed to limit subsiding and shifting during thaws. Some lessons might be drawn from recent strategies to off-set weather effects on the Trans-Alaska Pipeline.

- *Introduce proactive maintenance programs.* Routine monitoring, regular repairs, and strictly observed maintenance standards will be needed to ensure that preventable deterioration does not increase vulnerabilities.

- *Promote regional energy cooperation.* Trade and power swaps can help governments manage supply–demand constraints. Southeastern Europe is currently expanding regional grid interconnections, in what may be a promising trend.

- *Improve knowledge systems.* Progress in this area can provide more lead time and more accurate tracking of climate trends and weather events; data can be tailored for sector operations, maintenance, and design needs and for the development of workable emergency plans.

- *Provide a supporting framework for action.* The above initiatives could be supported through regulation, incentives for change, and, most important, outreach to key stakeholders.

Transport: Taking on Another Increment of Challenge[4]

More extreme heat, heavier precipitation, and periodic flooding carry implications for the planning, design, construction, and maintenance of transportation infrastructures. At the same time, on the demand side, weather conditions may change the ways that people use transportation.

The greatest concerns revolve around a cluster of extremes: rising sea levels, storm surges, heavier rainfall or snowstorms, and more days of intense heat. Coastal infrastructure on the Baltic and Black seas may require costly upgrading or may have to be moved altogether. With higher winds and more storms, railways, bridges, harbor structures, tunnels, and cranes in Central Europe and the Baltic coasts will become more vulnerable. More intense rains can stress transport systems, with pavement subgrades becoming less stable and retaining walls and abutments weakening. Flooding can lead to landslides and slope failure, washing out roads and railway lines. At the other extreme, long periods of intense heat or drought—as projected for much of Central Asia, Southeastern Europe, and the Caucasus—could lead to settling of soil beneath key structures and roads.

More extreme temperatures alone can accelerate road deterioration, particularly in Central Asia. In parts of Kazakhstan, the government already has imposed restrictions on truck travel to limit wear and tear during the scorching summer months when the asphalt softens. Elsewhere, changes in the freeze-thaw cycles can result in road damages. Specifically, degradation in the permafrost in northern and eastern Russia may affect a number of structures, including sections of the Trans Siberian Railway and airports serving remote communities.

Rural communities, already isolated and separated from some essential services, may become more marginalized if roads deteriorate or become impassable as a result of landslides or slope failures. Earth and gravel roads are easily damaged in heavy rainstorms; shorter, warmer winters shrink the length of time ice roads can be used. This is a critical issue for forestry and for oil and gas exploitation in Russia, where these sectors depend on ice-road travel.

Transportation planners and decision makers will face new challenges. Flooding and storm surges will affect multiple structures and systems across a wide area. At times, broader regional or cross-border cooperation will be required to solve a particular problem. Financial constraints will complicate and limit the planning process, particularly since climate change issues are not normally factored into budget plans.

Planners can fashion no-regret policies that generate direct or indirect benefits, significant enough to offset the immediate costs regardless of how extreme the climate change impacts turn out to be. Improved maintenance and rehabilitation programs prepare structures for climate-related stresses but are also good investments under any weather scenario. Meanwhile, governments can be encouraged to provide insurance against climate extremes, which can no longer be categorized as unforeseeable events. Public–private partnerships may help in providing this coverage.

A number of concrete actions will help to limit risks:

• Transportation agencies should establish systems for climate-attuned monitoring of key structures—for example, systems to measure bridge supports for the effects of heat stress or new pressures from changing water levels. Sensor technologies and computer processing advances make it possible to create more intelligent transportation systems that in effect track their own stress levels. Development of temperature-resistant materials will allow decision makers to make more optimal maintenance and rehabilitation choices.

- Planners can update design standards for key transport systems, incorporating current projections for warming, new precipitation patterns, and higher seas.

- New information and communications systems will have to ensure not only accurate and timely storm warnings and weather information, but also efficient communication of key information to transportation managers. More frequent intense storms will require the establishment of permanent evacuation routes and other emergency plans.

- Decision makers should acquire new technologies to help them understand and manage climate-related challenges. Digital elevation maps, satellite-based monitoring, and computer-assisted scenario planning could make a critical difference.

- Institutionalized mechanisms for knowledge sharing and communication between climate scientists and transportation professionals can help fill in the missing practical information decision makers need to identify and address the most vulnerable features of the larger transport system.

Notes

1. This section is based on "Achieving Urban Climate Adaptation in Europe and Central Asia" by JoAnn Carmin and Yan F. Zhang, a background paper prepared for this book.
2. This section is based on "Adapting to Climate Change in Europe and Central Asia; Background Paper on Water Supply and Sanitation" by Barbara Evans and Michael Webster, a background paper prepared for this book.
3. This section is based on "Europe and Central Asia Region: How Resilient Is the Energy Sector to Climate Change?" by Jane Ebinger, Bjorn Hamso, Franz Gerner, Antonio Lim, and Ana Plecas, a background paper prepared for this book.
4. This section is based on "Climate Change Adaptation in the Transport Sector" by Ziad Nakat, a background paper prepared for this book.

CHAPTER 7

Protection and Preparation: Disaster Risk Management and Weather Forecasting

Tim Carrington, Lucy Hancock,
Jolanta Kryspin-Watson, Sonja Nieuwejaar, John
Pollner, Marina Smetanina, and Vladimir Tsirkunov

Over the past 30 years, natural disasters have cost countries in Eastern Europe and Central Asia (ECA) about US$70 billion in economic losses (Pusch 2004). Most of the damage has occurred in Armenia, Poland, Romania, the Russian Federation, and Turkey. Meanwhile, climate change scenarios project even more frequent weather extremes, including increased flooding, heat waves, and drought, which will cause even greater losses. Changing trends and the impact of crossing temperature thresholds or tipping points (such as polar ice sheet collapse) can also set off abrupt disasters.

Taking steps to reduce the risks to people and structures from weather-related disasters is a worthwhile endeavor with high returns. By investing in strategies and systems for lowering the risk from one hazard, governments can strengthen a society's capacity to prepare for and adapt to a range of other threats. Planning for extremes will lessen physical damages and save lives while softening the economic impact. Climate change and the associated projected increases in

This chapter was drafted by Tim Carrington, based on a background paper prepared for this book, "Climate Change Adaptation in Europe and Central Asia: Disaster Risk Management" by John Pollner, Jolanta Kryspin-Watson, and Sonja Nieuwejaar; and "Weather and Climate Services in Europe and Central Asia" by Lucy Hancock, Vladimir Tsirkunov, and Marina Smetanina (World Bank Working Paper No. 151).

floods, heat waves, droughts, and snow emergencies only increase the importance of effective disaster risk management.

An essential starting point is to define risk management as a priority. From this base, a range of actions—from hazard warning and monitoring systems to financial instruments and disaster insurance products—can help countries manage hazards caused or intensified by climate change.

Monitoring the weather to know when extremes are coming is a critical public service, but it is one that has deteriorated through underfunding and other pressures that characterize the post-Soviet transition in most of the region. Information technologies have fallen behind, as has training for key personnel. While weather monitoring systems in the region have deteriorated, systems in other parts of the world have become more reliable. In many parts of the world, thanks to improved technology, seven-day forecasts are now nearly as accurate as three-day forecasts were in the early 1980s (Hancock, Tsirkunov, and Smetanina 2008).

Most countries in ECA have both the opportunity and the need to catch up with advances in weather forecasting and to employ improved systems for managing disaster risks. Sophisticated disaster risk management lessens a country's vulnerability to weather extremes; and improved weather tracking and forecasting helps anticipate emergencies and provide protection for human life and critical structures. By making the necessary (often modest) investments today, countries not only could contain losses from disasters but also provide a variety of useful services that would benefit productive sectors such as agriculture, aviation, and energy.

What follows is an analysis of current shortcomings in disaster management and weather forecasting in ECA, and the options that exist to better prepare for weather risks.

Softening the Blow When Disaster Strikes[1]

Countries need strategies to lessen the impacts of natural hazards and the environmental and structural breakdowns they cause. Some aspects of these strategies involve physical structures, while others focus on information systems or financial protection through insurance.

Analyzing the Current Capacity in ECA

In the difficult transition from centrally planned economies, the region has overhauled most political, social, and administrative structures, demilitarizing and restructuring many disaster management functions in the process. The restructuring and decentralization that occurred—carried out in an environment of systemic change and, in some countries, political instability—inevitably left gaps in the responsibility for maintaining and improving existing mechanisms and services. A 2004 study analyzed the capacities of all ECA countries to manage the multiple risks posed by natural disasters (Pusch 2004). In many Eastern European and Central Asian countries the existing mechanisms, which are insufficient for the current level of vulnerabilities, will be more inadequate still if the more extreme scenarios projected by climate change models materialize. Key weaknesses and areas of possible improvements include the following (adapted from Pusch 2004):

- *The concept of hazard risk management is not fully institutionalized.* Countries have elements of a new regulatory framework in place, but many governments lack statutory authority to devise and execute comprehensive, multi-sectoral disaster risk management programs.

- *Coordination mechanisms among authorities are underdeveloped.* Countries need better coordination among sectors, as well as stronger linkages between the central and local levels.

- *Hazard warning and monitoring systems require improvement.* Hydro-meteorological systems in the region need to incorporate recent technological advances that have dramatically strengthened forecasting capacities in other countries.

- *Economic considerations are not fully integrated in investment decisions.* Disaster risk management needs to incorporate rigorous cost-benefit or cost-effectiveness analyses so that investment priorities can be solidly established.

- *Catastrophe risk-financing tools are not fully used.* Most countries in the region can potentially access capital-market instruments to lessen the risks posed by natural disaster. But officials need expert support to master the available tools that other countries have already begun to use.

- *Funding of disaster risk mitigation is insufficient.* Recovery and reconstruction are much more costly in the aftermath of a disaster; shifting investments away from cleanup toward mitigation of risks can lower costs significantly.

- *Information and communication systems require upgrading.* Countries need the capacity to gather, interpret, and communicate vital information during an emergency. Some countries in the region, including Croatia, Romania, and Turkey, have initiated improvements in their emergency communication and information systems, but many others are lagging behind.

TABLE 7.1

**Reported Increased Incidence
of Weather-Induced Disasters in ECA**

Country	Hazard
Bulgaria	Cold wave, floods
Croatia	Floods
Czech Republic	Cold wave, floods
Estonia	Cold wave
Hungary	Windstorms, floods
Latvia	Snowfall, extreme cold, power shortage
Lithuania	Snowfall, extreme cold, power shortage
Moldova	Snowfall, extreme cold, power shortage
Montenegro	Floods
Poland	Cold wave, floods
Romania	Cold wave, floods
Russian Federation	Cold wave
Serbia	Floods
Slovak Republic	Floods
Turkey	Cold wave, floods

Source: Pollner, Kryspin-Watson, and Nieuwejaar 2008, based on data from
SIGMA—Swiss Re.

Evidence suggests that countries are already experiencing more frequent episodes of extreme weather. SIGMA, the catastrophe analysis arm of Swiss Re (one of the major global reinsurance companies), has also reported increasing incidences of weather-induced disasters in countries of the region (table 7.1).

With climate change contributing to the increase in weather extremes, disaster risk management becomes an urgent component of any climate change adaptation program. To reduce vulnerability, a disaster risk management program must incorporate five key elements:

- Risk assessment

- Risk mitigation investments addressing specific hazards

- Catastrophe risk financing

- Institutional capacity building

- Emergency preparedness and management.

FIGURE 7.1
Economic Loss Potential of Catastrophic Events for ECA Countries

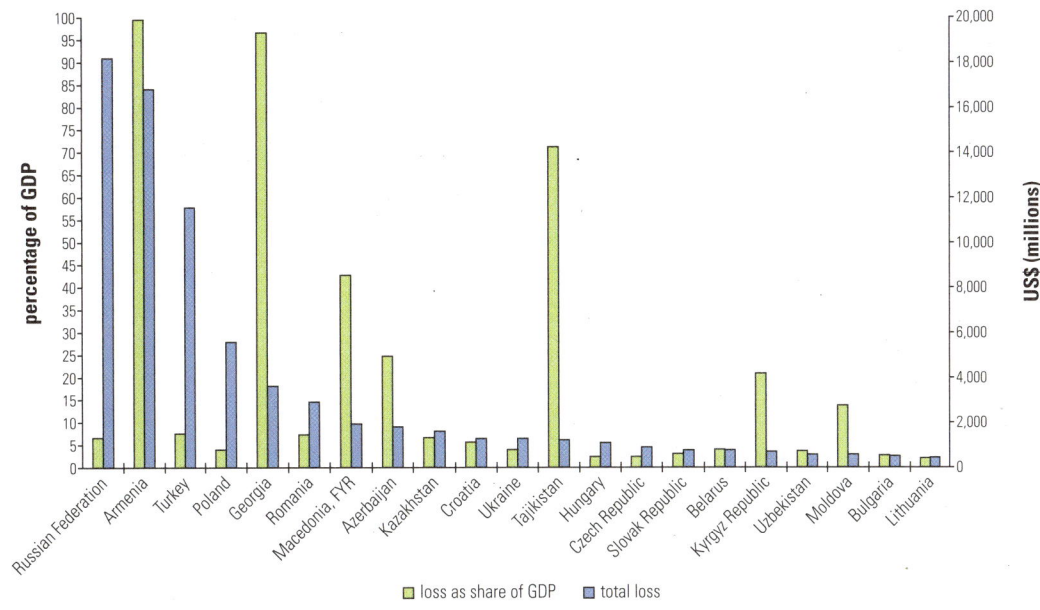

Source: Pusch 2004.

Note: Due to the structure of the original source data, this figure includes earthquakes, which are not climate-related, but excludes climate-sensitive droughts and forest fires, as well as industrial accident hazards.

What might seem like a low-probability event can translate into a major blow to the economy. Catastrophic events that have an annual probability of occurrence of 0.5 percent threaten an economic loss that exceeds 20 percent of gross domestic product (GDP) for Armenia, Azerbaijan, Georgia, the Former Yugoslav Republic of Macedonia, and Tajikistan; 10 percent in the Kyrgyz Republic and Moldova; and 5 percent in Kazakhstan, Romania, Russia, Southeastern Europe, and Turkey. Figure 7.1 shows the economic loss potential of catastrophic events for the GDP of many ECA countries.

Spreading the Risk: Budgeting, Facilitating, and Accessing Insurance Protection

For the most part, current government budgets in ECA are grossly insufficient to finance large losses from extreme events, while insurance protection is largely inadequate to make up for the shortfall. An exception in the region is the Czech Republic. Flooding in 2002 caused US$3 billion in damages—equivalent to nearly 4 percent of GDP. However, nearly 40 percent of these losses in 2002 were insured as a result of an increase in demand for insurance that followed the

1997 floods (CEA Insurers of Europe 2007). But that still left 60 percent of the losses, or more than 2 percent of GDP, uninsured—much of it public infrastructure. A rational fiscal policy would budget annual premiums for insurance, avoiding the greater disruption of having to make massive expenditures once a disaster hits.

Within the global catastrophe insurance market, insurance premiums for extreme events fluctuate, complicating budget planning for government. However, vulnerable countries can protect themselves against catastrophic risk *and* premium volatility by using capital markets. The annualized risk of extreme losses from weather events induced by climate change to date has been in the 1 percent range, a level of risk that is normally acceptable to the markets. Thus, there is room for the broader private financial sector to absorb and spread the risks, both domestically and internationally. Two potentially useful mechanisms for more efficient management of catastrophic risk are pooled insurance coverage supported by liquidity and credit enhancement facilities and weather-indexed bonds to securitize risk.

Multilateral development institutions can support the development of these mechanisms, while still ensuring actuarially fair premiums. For example, the Caribbean Catastrophe Risk Insurance Facility implemented a risk pool with World Bank support, which reduced the cost of premiums paid by the island governments for coverage for extreme hurricane and earthquake events. The World Bank also assisted Mexico in launching an indexed catastrophe bond for coverage in the event of a massive earthquake. Currently, the World Bank is assisting a number of ECA countries in establishing a Catastrophe Risk Insurance Facility, which will pool individual disaster risks and provide coverage to homeowners and businesses in Southeastern and Central Europe.

Reinsurance from foreign companies can lower the price of disaster insurance, but the reinsurance itself can be costly. When the domestic insurers shift all but very low levels of risk to reinsurance companies abroad, the coverage is generally expensive because of the high likelihood that it will be triggered. Contracting reinsurance only for much higher levels of loss lowers the premiums. But when catastrophic events occur, and reinsurance companies experience massive payouts, premiums rise and extreme-event reinsurance markets then tighten. Thus, while helpful, reinsurance is not a panacea.

When global insurance and reinsurance markets become too costly, an alternative is the catastrophe bond market, which exists in Europe, Japan, and the United States. Investors buy high-yield bonds from the party that seeks to be insured. These bonds can either be backed by premiums collected on insured assets or be structured as a financial option using other calibrations. Many of these risk management methods could be adapted for use in ECA countries.

Direct government involvement can play a part in insuring against losses from extreme events associated with climate change, with catastrophe bonds or reinsurance arrangements available as options. Another insurance innovation that national governments could facilitate is the creation of a central fund for catastrophe risks. A mechanism could be established in which liquidity and credit enhancement facilities support insurance coverage against catastrophic risks. The domestic insurance industry would transfer catastrophic coverage to a central fund regulated by the government but operated by the insurance industry itself. The risks covered would not be reflected on the balance sheets of local insurers but would be liabilities of the pooled fund. The international insurance industry could then reinsure climate-induced catastrophic coverage under the fund up to a specified loss limit. Multilateral institutions might provide contingent credit at the next highest loss level, supporting the liquidity of the fund in the event of immediate large losses in the initial years of operation.

Finally, weather-indexed bonds are another insurance instrument that can mitigate climate-related risk. Catastrophe bonds based on payouts linked to measurable weather events (as reflected in weather indexes or parametric measures) have the advantage of being relatively easy to implement once a reliable weather measurement mechanism is identified. They bypass the traditional insurance loss adjustment process, which requires site-by-site evaluation of losses before indemnity is provided. The payout is simply based on the weather index reaching a certain range.

The main risk with weather-indexed instruments is that the payout is not directly linked to actual losses. A payment might be made—with the bondholder losing interest and principal—even though the insured party experiences no loss. Alternately, the insured party may experience a loss but receive no indemnity because the parametric index was not triggered. Still the instrument might be attractive to international investors for portfolio diversification, since natural disasters have little or no correlation with global financial market trends.

Mitigating the Risks

Insurance schemes help countries spread the costs of a disaster brought on by extreme weather. But it is also important to take steps that actually mitigate the risks by making structures, people, and ecological systems less vulnerable to damage from extreme weather. Following are ways—several already alluded to in previous chapters—that governments can lower the risks (presented in greater detail in the background paper by Pollner et al.):

- *Retrofitting.* Modifying existing structures to help them withstand natural disasters. Examples include installing backup valves in sewage and water pipes, elevating structures, and installing storm shutters or foundation strengthening.

- *Regulations.* By controlling the use of land and the construction of buildings, governments can significantly reduce the potential losses from disasters. In some cases, risks could be lowered simply by enforcing existing zoning and building codes.

- *Protective structures.* Structures such as sea walls and levees can protect buildings and people and mitigate the impact of floods and storms.

- *Natural resource management.* Better management of natural resources—controlling erosion, protecting forests, and restoring wetlands—preserves ecosystem services that help reduce the consequences of weather shocks.

A critical element in reducing vulnerability is an analysis of human settlements and infrastructure in high-risk areas. Geographic Information Systems, with layers of digital data, can be used to create risk maps and data resources that help decision makers to assess and locate risks, take preventive and preparedness measures, and set investments priorities.

Some risk mitigation steps will need to be specific to particular hazards. Fire protection is an important component in protecting forest and grassland, particularly in Southeastern Europe, where the growing frequency of wildfires highlights the dangers. Particularly helpful might be the development of an early warning system to predict when and where forest fires are most likely to occur, as well as a monitoring system that helps with response coordination.

Understanding When Extreme Weather Is Coming[2]

Thirty years ago, weather forecasting and overall hydrometeorological (hydromet) services in many ECA countries were among the most advanced and reliable in the world. However, the status of most weather services among the ECA countries has deteriorated considerably in the last two decades, mainly as a consequence of persistent underfinancing of services during the transition that followed the end of central planning.

Performance has deteriorated in virtually all the region's weather services, and certain agencies are on the brink of collapse. Many surface data collection stations have closed, and those that remain open record a more limited set of parameters on a less frequent basis using instruments that are aging and failing. Communications equipment to convey station data to headquarters for analysis is often obsolete, unreliable, and labor-intensive. Training is inadequate both to keep the skills of senior staff current and to prepare a sufficient number of qualified incoming staff.

Worrisome examples of shortcomings proliferate in ECA. Turkmenistan has no upper-atmosphere sensing stations at present, which compromises the safety of aviation in Ashgabat. Tajikistan's network of weather stations was severely damaged in the 1992–98 conflict, and reliable weather time series are generally unavailable. Kazakhstan does not have meteorological radars or specialized stations to receive satellite data. In Georgia, most meteorological and hydrological stations have closed, upper-air observations have halted, and only one meteorological radar is in operation. In Ukraine, 90 percent of all instruments have exceeded their intended service life, and many facilities are in urgent need of repair.

In sum, the range of the accumulated problems is so great that, without massive modernization, networks in a number of ECA countries are on their way to becoming completely dysfunctional. No longer able to count on their own weather services, countries would be forced to depend on low-resolution forecasts prepared by others that would miss significant local and rapid-onset hazards, including floods, frosts, and severe storms. The perils of a weakening forecast capacity have become evident in Russia where the share of hazardous weather phenomena that were not picked up and forecast increased from 6 percent at the beginning of 1990s to 23 percent only 10 years later.

Recent research underscores the value of investment in hydromet services. A study in China concluded that expenditures on the meteorological service had a cost-benefit ratio of between 1 to 35 and 1 to 40 (Guocai and Wang 2003). An estimate in Mozambique suggested a cost-benefit ratio of 1 to 70 for investment in the meteorological service, which needed to be rebuilt after that country's civil war. Mozambique saw directly the consequences of being uninformed and unprepared: when floods swept the country in 2000, it cost Mozambique 12 percent of its GDP in direct and indirect costs (World Bank 2001). A number of easily accessible technologies and available upgrades to weather forecasting systems could indeed be affordable, as long as governments budget for, staff, and equip hydromet services at adequate levels. Some examples include the following:

- *Bandwidth.* A global telecommunications system organized by the United Nations World Meteorological Organization (WMO) shares global forecasts and data. Yet, ECA's underfunded agencies are often unable to make full use of this resource for lack of bandwidth to download large files.

- *Satellite dishes.* Weather satellites launched over Europe broadcast low-cost, or no-cost, images of storm systems, fires, coastal zone pollution, and other environmental data. However, many weather agencies in ECA cannot make use of this critical data because they lack satellite dishes or processing capacity.

- *Local area modeling.* Global communities of experts have jointly devised open-source models for weather prediction that lend themselves to local weather forecasting and can be run on computers only slightly more powerful than commonly used desktops. Many countries would benefit from training in use of these packages.

- *Forecasting workstations.* In some countries, satellite and radar data from neighboring countries may be available, but weather agencies often lack the workstations and software to make use of the data for forecasting purposes.

These widely available tools will not take the place of the more comprehensive modernization that most of ECA's hydromet systems need. To manage more frequent weather extremes and changing patterns in heat and precipitation, national systems will need to draw on data from radars, surface weather stations, upper-air sounding stations, hydrological stations, and specialized networks. These inputs need to flow to a national headquarters through efficient telecommunications networks. Staff require training to produce accurate forecasts covering a three-day period, along with useful seven-day forecasts specific to locations within 10 kilometers. This level of performance not only would help countries to warn citizens of pending weather catastrophes, but also would provide economically valuable information to the agriculture, water management, and transport sectors.

Often, the benefits of timely and accurate forecasts, both for reducing disaster impacts (box 7.1) and improving decision making in agriculture, can be easily measured. Increased accuracy in forecasting would assist in the timing of irrigation and fertilizer application and

BOX 7.1

Poland's Flood Disaster Leads to Stepped-up Preparation

Poland, caught by surprise in massive floods in 1997, resolved to be better prepared to face future weather extremes. A Flood Emergency Project, supported by the World Bank and the European Bank for Reconstruction and Development, included development of a monitoring, forecasting, and warning system; flood prevention planning; and upgrading of flood prevention infrastructure. It also supported the development of non-structural measures to limit damage, including regulations on the economic use of risky areas, flood-impact minimization plans prepared by local communities and groups, warning systems, and flood insurance.

The upgraded system cost US$62 million to establish and US$8 million per year to maintain. The investment is small when set against the costs of the 1997 disaster: the floods inundated dozens of cities and hundreds of villages, costing 55 people their lives and causing US$3.4 billion in damages.

Source: Hancock, Tsirkunov, and Smetanina 2008.

in pest and disease control, avoiding over-application that raises input costs and exacerbates environmental damage. There is abundant evidence that farmers in Albania, Montenegro, Tajikistan, and Uzbekistan would benefit significantly from improved monitoring and forecasting.

Forecasts also would enable mitigation of frost damage, which is a serious problem for agriculture in Armenia, Bosnia and Herzegovina, Kazakhstan, FYR Macedonia, Moldova, Montenegro, Ukraine, and Turkmenistan, among others. Approaches to mitigate the effects of sudden freezes are being developed globally, but cost-effective application depends on accurate forecasting.

Extreme weather does not respect national borders. Countries in the region have unnecessarily suffered because critical weather information was not shared effectively among neighboring countries. Damaging weather patterns of special importance include Atlantic and Mediterranean cyclones and intrusions of cold air from the far north. Rapidly changing, dangerous events are best monitored through transboundary data sharing that goes beyond WMO requirements (Ogonesyan 2004). However, gaps in data sharing persist in ECA, often because of political instability and conflict. The Caucasus region and the Western Balkans have experienced significant breaks and gaps in data sharing as a result of the clashes and upheavals of the past two decades.

Other parts of the world have had some success with regional multi-hazard centers and public-private partnerships, where private firms play a role in processing or disseminating weather data. Both these approaches could prove useful in rebuilding ECA's forecasting capacity.

Different subregions within ECA face different challenges in upgrading their systems. In the mountainous Balkans, there is a relatively sparse network of weather stations, limiting countries' ability to update and localize global weather data. Accessing and incorporating data from Greek weather stations would be helpful. Similarly, the Caucasus region, which suffers from a paucity of operating weather stations, might benefit from heavier use of data from Turkey.

In the water-stressed, mountainous areas of Central Asia and the Caucasus, weather systems are especially critical for water management. In these countries, where snow and glacier melt feeds local rivers, monitoring of snow accumulation and glacier volume is needed to project water resources and water quality. Central Asia's mountains pose other challenges. Oceanic air masses moving over them can first appear to exhaust their supply of moisture, but then rise and cool, collecting sufficient moisture to cause heavy rains and flash flooding in Kazakhstan, the Kyrgyz Republic, Tajikistan, and Uzbekistan. Additional monitoring stations would help forecasters track the changing air patterns.

Conclusion

The majority of ECA countries are not among the world's most vulnerable to weather extremes. According to data assembled from ministries responsible for emergencies in a number of ECA countries, annual losses from weather events range from 0.5 percent to 1.9 percent of GDP. This compares to a global range of 0.1 percent to 5 percent (Hancock, Tsirkunov, and Smetanina 2008). However, unusually large storms and floods can cause, and have caused, far higher damages. As weather extremes become more frequent, it makes sense to act early to minimize the losses.

Investment in forecasting systems that provide reliable and timely warning is critical, with multiple analyses demonstrating that investments to modernize hydromet systems pay for themselves many times over. Equally critical are disaster risk management measures, to lessen physical exposure to weather-related disasters as well as to limit or transfer economic losses when disasters do occur.

Another critical concern is to provide adequate safety nets for those who, despite improved warning systems and disaster risk management plans, suffer devastating losses in disasters. Analysis of the adequacy of existing safety net programs, and recommendations for practical ways of filling existing gaps, will be a high priority for ECA countries and their international partners given the additional strains of a changing climate.

Notes

1. This section is based on "Climate Change Adaptation in Europe and Central Asia: Disaster Risk Management" by John Pollner, Jolanta Kryspin-Watson, and Sonja Nieuwejaar, a background paper prepared for this book.
2. This section is based on Hancock, Tsirkunov, and Smetanina (2008).

References

Adger, W. N., S. Agrawala, M. M. Q. Mirza, C. Conde, K. O'Brien, J. Pulhin, R. Pulwarty, B. Smit, and K. Takahashi. 2007. "Assessment of Adaptation Practices, Options, Constraints and Capacity." In *Climate Change 2007: Impacts, Adaptation, and Vulnerability—Contribution of Working Group II to the Fourth Assessment Report of the Intergovernmental Panel on Climate Change,* ed. M. L. Parry, O. F. Canziani, J. P. Palutikof, P. J. van der Linden, and C. E. Hanson, 717–44. Cambridge, U.K.: Cambridge University Press.

Agaltseva, N. 2008. "Prospective Change of the Central Asian Rivers Runoff with Glaciers Feeding under Different Climate Scenarios." Research Hydrometeorological Institute, Uzhidromet, Uzbekistan. In *Geophysical Research Abstracts* 10: EGU2008-A-00464, 2008SRef-ID: 1607-7962/gra/EGU2008-A-00464. EGU General Assembly 2008.

Alam, A., M. Murthi, R. Yemstov, E. Murrugarra, N. Dudwick, E. Hamilton, and E. Tiongson, eds. 2005. *Growth, Poverty, and Inequality: Eastern Europe and the Former Soviet Union.* Washington, DC: World Bank.

Alamanov, S. K., V. M. Lelevkin, O. A. Podrezov, and A. Podrezov. 2006. "Climate Change and Water Problems in Central Asia: Learning Course for Students." [In Russian.] World Wildlife Fund–Russia and United Nations Environment Programme, Moscow and Bishkek.

Alcamo, J., J. M. Moreno, B. Nováky, M. Bindi, R. Corobov, R. J. N. Devoy, C. Giannakopoulos, E. Martin, J. E. Olesen, and A. Shvidenko. 2007. "Europe." In *Climate Change 2007: Impacts, Adaptation, and Vulnerability—Contribution of Working Group II to the Fourth Assessment Report of the Intergovernmental Panel on Climate Change*, ed. M. L. Parry, O. F. Canziani, J. P. Palutikof, P. J. van der Linden, and C. E. Hanson, 541–80. Cambridge, U.K.: Cambridge University Press.

Alexandrov, V. A. 1997. "Vulnerability of Agronomic Systems in Bulgaria."
    Climatic Change 36 (1–2): 135–49.

Anisimov, O., and S. Reneva. 2006. "Permafrost and Changing Climate: The
    Russian Perspective." *Ambio* 35 (4): 169–75.

Anisimov, O., and D. Vaughan. 2007. "Polar Regions." In *Climate Change 2007:
    Impacts, Adaptation, and Vulnerability—Contribution of Working Group II to the
    Fourth Assessment Report of the Intergovernmental Panel on Climate Change*, ed.
    M. L. Parry, O. F. Canziani, J. P. Palutikof, P. J. van der Linden, and C. E.
    Hanson, 653–86. Cambridge, U.K.: Cambridge University Press.

Antypa, A., M. Pouyiouka, E. Spanos, K. Petrochilou, V. Karabassi, K.
    Paraskeva, N. Alexandropoulos, and M. Toutouza. 2007. "Tuberculosis
    among Immigrants: Our Experience." *International Journal of Antimicrobial
    Agents* 29: 454.

Araujo, Miguel B., and Carsten Rahbek. 2006. "How Does Climate Change
    Affect Biodiversity?" *Science* 313: 1396–97.

Arnell, N. W., and E. K. Delaney. 2006. "Adapting to Climate Change: Public
    Water Supply in England and Wales." *Climatic Change* 78 (2–4): 227–55.

Australian Government. 2005. "Climate Change Risk and Vulnerability: Pro-
    moting an Efficient Adaptation Response in Australia." Report prepared
    by the Allen Consulting Group for the Australian Greenhouse Office,
    Department of the Environment and Heritage, Canberra.

———. 2006. "Climate Change Scenarios for Initial Assessment of Risk in
    Accordance with Risk Management Guidance." Report prepared by K.
    Hennessy, I. Macadam, and P. Whetton of the Commonwealth Scientific
    and Industrial Research Organisation (CSIRO) for the Australian Green-
    house Office, Department of the Environment and Heritage, Canberra.

Azar, C., and K. Lindgren. 2003. "Catastrophic Events and Stochastic Cost-
    Benefit Analysis of Climate Change." *Climatic Change* 56 (3): 245–55.

Baettig, M. B., M. Wild, and D. M. Imboden. 2007. "A Climate Change Index:
    Where Climate Change May Be Most Prominent in the 21st Century."
    Geophysical Research Letters 34.

Bari, A., B. Grbec, and D. Bogner. 2008. "Potential Implications of Sea-Level
    Rise for Croatia." *Journal of Coastal Research* 24 (2): 299–305.

Bates, B. C., Z. W. Kundzewicz, S. Wu, and J. P. Palutikof, eds. 2008. "Climate
    Change and Water." Technical Paper of the Intergovernmental Panel on
    Climate Change (IPCC), IPCC Secretariat, Geneva.

BEEN (Baltic Energy Efficiency Network for the Building Stock). 2007. *BEEN
    Project Results, Including Detailed Findings and Recommendations—Practical
    Manual*. Berlin: Berlin Senate Department for Urban Development.

Beniston, M., D. B. Stephenson, O. B. Christensen, C. A. T. Ferro, C. Frei,
    S. Goyette, K. Halsnaes, T. Holt, K. Jylhä, B. Koffi, J. Palutiko, R. Schöll,
    T. Semmler, and K. Woth. 2007. "Future Extreme Events in European
    Climate: An Exploration of Regional Climate Model Projections." *Climatic
    Change* 81 (Suppl. 1): 71–95.

Bianchi, T. S., P. Westman, and C. Rolff. 2000. "Cyanobacterial Blooms in the
    Baltic Sea: Natural or Human Induced?" *Limnology and Oceanography* 45
    (3): 716–26.

Brylski, P., and S. Abdulin. 2003. *Biodiversity Strategy for the Europe and Central Asia Region*. Washington, DC: World Bank.

Bukovska, B. 2002. "Difference and Indifference: Bringing Czech Roma Ghettoes to Europe's Court." *European Union Monitoring and Advocacy Program (EUMAP) Online Journal*. http://www.eumap.org/journal/features/2002/may02/czechromaghettoes.

Burton, I., and B. Lim. 2005. "Achieving Adequate Adaptation in Agriculture." *Climatic Change* 70: 191–200.

Carmin, J., and Y. F. Zhang. 2008. "Achieving Urban Climate Adaptation in Europe and Central Asia." Background paper prepared for report, World Bank, Washington, DC.

Carter, T. R., R. N. Jones, X. Lu, S. Bhadwal, C. Conde, L. O. Mearns, B. C. O'Neill, M. D. A. Rounsevell, and M. B. Zurek. 2007. "New Assessment Methods and the Characterisation of Future Conditions." In *Climate Change 2007: Impacts, Adaptation, and Vulnerability—Contribution of Working Group II to the Fourth Assessment Report of the Intergovernmental Panel on Climate Change*, ed. M. L. Parry, O. F. Canziani, J. P. Palutikof, P. J. van der Linden, and C. E. Hanson, 133–72. Cambridge, U.K.: Cambridge University Press.

Cavazos, T., and B. C. Hewitson. 2005. "Performance of NCEP-NCAR Reanalysis Variables in Statistical Downscaling of Daily Precipitation." *Climate Research* 28: 95–107.

CEA Insurers of Europe. 2007. "Reducing the Social and Economic Impact of Climate Change and Natural Catastrophes: Insurance Solutions and Public-Private Partnerships." CEA Insurers of Europe, Brussels.

Cenacchi, N. 2008a. "Adaptation to Climate Change in Coastal Areas of the ECA Region." Background paper prepared for report, World Bank, Washington, DC.

———. 2008b. "Biodiversity Adaptation to Climate Change in the ECA Region." Background paper prepared for report, World Bank, Washington, DC.

Chichilnisky, G. 2000. "An Axiomatic Approach to Choice under Uncertainty with Catastrophic Risks." *Resource and Energy Economics* 22 (3): 221–31.

Christensen, J. H., T. R. Carter, M. Rummukainen, and G. Amanatidis. 2007. "Evaluating the Performance and Utility of Regional Climate Models: The PRUDENCE Project." *Climatic Change* 81 (Suppl. 1): 1–6.

Chorus, Ingrid, and Jamie Bartram. 1999. *Toxic Cyanobacteria in Water: A Guide to Their Public Health Consequences, Monitoring and Management*. Geneva: World Health Organization.

Clarke, S., J. Kersey, E. Trevorrow, R. Wilby, S. Shackley, J. Turnpenny, A. Wright, A. Hunt, and D. Crichton. 2002. *London's Warming: The Impacts of Climate Change on London. Technical Report*. London, U.K.: London Climate Change Partnership.

Cline, W. R. 2007. *Global Warming and Agriculture: Impact Estimates by Country*. Washington, DC: Center for Global Development and Peterson Institute for International Economics.

Cruz, R. V., H. Harasawa, M. Lal, S. Wu, Y. Anokhin, B. Punsalmaa, Y. Honda, M. Jafari, C. Li, and N. Huu Ninh. 2007. "Asia." In *Climate Change 2007:*

Impacts, Adaptation, and Vulnerability—Contribution of Working Group II to the Fourth Assessment Report of the Intergovernmental Panel on Climate Change, ed. M. L. Parry, O. F. Canziani, J. P. Palutikof, P. J. van der Lindne, and C. E. Hanson, 469–506. Cambridge, U.K.: Cambridge University Press.

Csaki, C., H. Kray, and S. Zorya. 2006. "The Agrarian Economies of Central-Eastern Europe and the Commonwealth of Independent States: An Update on Status and Progress in 2005." Environmentally and Socially Sustainable Development Working Paper 46, World Bank, Washington, DC.

Dai, A., K. E. Trenberth, and T. Qian. 2004. "A Global Data Set of Palmer Drought Severity Index for 1870–2002: Relationship with Soil Moisture and Effects of Surface Warming." *Journal of Hydrometeorology* 5: 1117–30.

Dilley, M., R. S. Chen, U. Deichmann, A. L. Lerner-Lam, and M. Arnold. 2005. *Natural Disaster Hotspots: A Global Risk Analysis*. Washington, DC: World Bank.

Dronin, N., and A. Kirilenko. 2008. "Climate Change and Food Stress in Russia: What If the Market Transforms as It Did during the Past Century?" *Climatic Change* 86: 123–50.

Easterling, W. E., P. K. Aggarwal, P. Batima, K. M. Brander, L. Erda, S. M. Howden, A. Kirilenko, J. Morton, J.-F. Soussana, J. Schmidhuber, and F. N. Tubiello. 2007. "Food, Fibre, and Forest Products." In *Climate Change 2007: Impacts, Adaptation, and Vulnerability—Contribution of Working Group II to the Fourth Assessment Report of the Intergovernmental Panel on Climate Change*, ed. M. L. Parry, O. F. Canziani, J. P. Palutikof, P. J. van der Linden, and C. E. Hanson, 273–314. Cambridge, U.K.: Cambridge University Press.

Ebinger, J., B. Hamso, F. Gerner, A. Lim, and A. Plecas. 2008. "Europe and Central Asia Region: How Resilient Is the Energy Sector to Climate Change?" Background paper prepared for report, World Bank, Washington, DC.

EEA (European Environment Agency). 2005. "Vulnerability and Adaptation to Climate Change in Europe." EEA Technical Report 7/2005, EEA, Copenhagen.

———. 2007. "Climate Change: The Cost of Inaction and the Cost of Adaptation." EEA Technical Report 13/2007, EEA, Copenhagen.

Elguindi, N., and F. Giorgi. 2007. "Projected Changes in the Caspian Sea Level for the 21st Century Based on AOGCM and RCM Simulations." In *Geophysical Research Abstracts* 9.

EM-DAT (Emergency Events Database). 2008. Centre for Research on the Epidemiology of Disasters (CRED), Université Catholique de Louvain, Louvain, France. http://www.emdat.be/Database/terms.html.

Environment Canada. 2004. "Threats to Water Availability in Canada." Scientific Assessment Report 3, Prowse and ASCD Science Assessments Series 1, National Water Research Institute, Burlington, ON.

European Commission. 2000. "Global Land Cover 2000." Joint Research Center, Institute for Environment and Sustainability. http://ies.jrc.ec.europa.eu/global-land-cover-2000.

———. 2005. "Natura 2000—Europe's Nature for You." Office for Official Publications of the European Communities, Luxembourg.

———. 2007. "Adapting to Climate Change in Europe—Options for EU Action." Commission staff working document accompanying the Green Paper, SEC (2007) 849, European Commission, Brussels.

Evans, B., and M. Webster. 2008. "Adapting to Climate Change in Europe and Central Asia: Background Paper on Water Supply and Sanitation." Background paper prepared for report, World Bank, Washington, DC.

FAO (Food and Agriculture Organization). 2006a. *Livestock's Long Shadow: Environmental Issues and Options*. Rome: FAO.

———. 2006b. "Country Profiles." In *FAO Statistical Yearbook*, vol. 1–2. Rome: FAO.

FAO (Food and Agriculture Organization) and EBRD (European Bank for Reconstruction and Development). 2008. "Fighting Food Inflation through Sustainable Investment: Grain Production and Export Potential in CIS Countries and Rising Food Prices—Causes, Consequences, and Policy Responses." FAO and EBRD, London.

Fay, M., and H. Patel. 2008. "A Simple Index of Vulnerability to Climate Change." Background paper prepared for report, World Bank, Washington, DC.

Fink, A. H., T. Brücher, A. Krüger, G. C. Leckebusch, J. G. Pinto, and U. Ulbrich. 2004. "The 2003 European Summer Heat Waves and Drought—Synoptic Diagnosis and Impact." *Weather* 59 (8): 209–16.

Finnish Environment Institute. 2007. *Assessing the Adaptive Capacity of the Finnish Environment and Society under a Changing Climate: FINADAPT—Summary for Policy Makers*, ed. Timothy Carter. Helsinki: Finnish Environment Institute.

Frolov, A. 2000. "Black and Caspian Seas Vulnerability and Adaptation: Implications of Accelerated Sea-Level Rise (ASLR) for Russia." Proceedings of the Synthesis and Upscaling of Sea-Level Rise Vulnerability Assessment Studies (SURVAS) Expert Workshop on European Vulnerability and Adaptation to Impacts of Accelerated Sea-Level Rise, Hamburg, Germany.

Füssel, H.-M. 2007. "Vulnerability: A Generally Applicable Conceptual Framework for Climate Change Research." *Global Environmental Change* 17 (2): 155–67.

Gagnon-Lebrun, F., and S. Agrawala. 2006. "Progress on Adaptation to Climate Change in Developed Countries: An Analysis of Broad Trends." Environment Directorate, Organisation for Economic Co-operation and Development, Paris.

GEF (Global Environmental Facility). 2002. *Transboundary Diagnostic Analysis for the Caspian Sea*. Baku: Caspian Environment Programme.

———. 2007. *Black Sea Transboundary Diagnostic Analysis*. Baku: Commission on the Protection of the Black Sea against Pollution.

Gill, S.E., J. F. Handley, A. R. Ennas, and S. Paul. 2007. "Adapting Cities for Climate Change: The Role of the Green Infrastructure." *Built Environment* 33 (1): 122–23.

Glantz, M. H., and I. S. Zonn. 2005. *The Aral Sea: Water, Climate and Environmental Change in Central Asia*. Geneva: World Meteorological Organization.

Groves, D. G., D. Knopman, R. J. Lempert, S. H. Berry, and L. Wainfan. 2008. "Presenting Uncertainty about Climate Change to Water-Resource Managers: A Summary of Workshops with the Inland Empire Utilities Agency." Technical Report of the Environment, Energy, and Economic Development Program, RAND Corporation, Santa Monica.

Guocai, Z., and H. Wang. 2003. "Evaluating the Benefits of Meteorological Services in China." *WMO Bulletin* 52 (4): 383–87.

Gushulak, B. D., and W. D. MacPherson. 2006. *Migration Medicine and Health: Principles and Practice.* Hamilton, Ontario: BC Decker.

Hajat, S., K. L. Ebi, R. S. Kovats, B. Menne, S. Edwards, and A. Haines. 2003. "The Human Health Consequences of Flooding in Europe and the Implications for Public Health: A Review of the Literature." *Applied Environmental Science and Public Health* 1: 13–21.

Hamers, F. F., I. Devaux, J. Alix, and A. Nardone. 2006. "HIV/Aids in Europe: Trends and EU-wide priorities." *Eurosurveillance* 11 (47). http://www.eurosurveillance.org/ViewArticle.aspx?ArticleId=3083.

Hancock, L., V. Tsirkunov, and M. Smetanina. 2008. "Weather and Climate Services in Europe and Central Asia: A Regional Review." Working Paper 151, World Bank, Washington, DC.

Hannah, L., G. F. Midgley, T. Lovejoy, W. J. Bond, M. Bush, J. C. Lovett, D. Scott, and F. I. Woodward. 2002. "Conservation of Biodiversity in a Changing Climate." *Conservation Biology* 16 (1): 264–68.

Hansen, J. W., S. Marx, and E. Weber. 2004. "The Role of Climate Perceptions, Expectations, and Forecasts in Farmer Decision-Making: The Argentine Pampas and South Florida." IRI Technical Report 04–1, International Research Institute for Climate Prediction, Palisades, NY.

Heinz Center. 2007. *A Survey of Climate Change Adaptation Planning.* Washington, DC: H. John Heinz III Center for Science, Economics, and the Environment.

HELCOM (Helsinki Commission). 2007. "Climate Change in the Baltic Sea Area: HELCOM Thematic Assessment in 2007." Baltic Sea Environment Proceedings 111, HELCOM and Baltic Marine Environment Protection Commission, Helsinki.

Helms, M., B. Büchele, U. Merkel, and J. Ihringer. 2002. "Statistical Analysis of the Flood Situation and Assessment of the Impact of Diking Measures along the Elbe (Labe) River." *Journal of Hydrology* 267: 94–114.

Holman, I. P., M. D. A. Rounsevell, S. Shackley, P. A. Harrison, R. J. Nicholls, P. M. Berry, and E. Audsley. 2005. "A Regional, Multi-sectoral and Integrated Assessment of the Impacts of Climate and Socio-economic Change in the U.K.: I Methodology." *Climatic Change* 71: 9–41.

Hovsepyan, A., and H. Melkonyan. 2007. "Model Simulations of Climate Change over the South Caucasus during the 21st Century." Armstatehydromet, Yerevan.

HSE (U.K. Health and Safety Executive). 2001. "Marine Risk Assessment." Offshore Technical Report 2001/063, HSE, London.

———. 2006. "Guidance on Risk Assessment for Offshore Installations." Offshore Information Sheet 3/2006, HSE, London.

IOM (International Organization of Migration). 2007. "Migration as It Is: An Overview of Migration in the Republic of Moldova." Chisinau.

IPCC (Intergovernmental Panel on Climate Change). 2007a. "Appendix I: Glossary." In *Climate Change 2007: Impacts, Adaptation, and Vulnerability. Contribution of Working Group II to the Fourth Assessment Report of the Intergovernmental Panel on Climate Change*, ed. M. L. Parry, O. F. Canziani, J. P. Palutikof, P. J. van der Linden, and C. E. Hanson, 869–84. Cambridge, U.K.: Cambridge University Press.

———. 2007b. "Summary for Policy Makers." In *Climate Change 2007: Impacts, Adaptation, and Vulnerability. Contribution of Working Group II to the Fourth Assessment Report of the Intergovernmental Panel on Climate Change*, ed. M. L. Parry, O. F. Canziani, J. P. Palutikof, P. J. van der Linden, and C. E. Hanson, 7–22. Cambridge, U.K.: Cambridge University Press.

———. 2007c. *Climate Change 2007: Synthesis Report. Contribution of Working Groups I, II, and III to the Fourth Assessment Report of the Intergovernmental Panel on Climate Change.* Geneva: IPCC.

Karaca, M., and R. J. Nicholls. 2008. "Potential Implications of Accelerated Sea-Level Rise for Turkey." *Journal of Coastal Research* 24 (2): 288–98.

Kattsov, V., G. A. Alekseev, T. V. Pavlova, P. V. Sporyshev, R. V. Bekryaev, and V. A. Govorkova. 2007. "Modeling the Evolution of the World Ocean Ice Cover in the 20th and 21st Centuries." *Izvestiya, Atmospheric and Ocean Physics* 43: 165–81.

Kattsov, V., V. Govorkova, V. Meleshko, T. Pavlova, and I. Shkolnik. 2008. "Climate Change Projections and Impacts in Russian Federation and Central Asia States." Background paper prepared for World Bank report, Voeikov Main Geophysical Observatory, St. Petersburg, Russia, and World Bank, Washington, DC.

Kaufman, D., A. Kraay, and M. Mastruzzi. 2008. Worldwide Governance Indicators (WGI) Database. World Bank, Washington, DC. http://info.worldbank.org/governance/wgi/.

Klein, R. 2008. "Mainstreaming Climate Adaptation into Development." Stockholm Environment Institute briefing note for the European Parliament Temporary Committee on Climate Change, Stockholm, July.

Kokorin, A. 2008. "Expected Impact of the Changing Climate on Russia and Central Asia Countries and Ongoing or Planned Adaptation Efforts and Strategies in Russia and Central Asia Countries." Background paper prepared for World Bank report, WWF (World Wide Fund for Nature) Russia, Moscow, and World Bank, Washington, DC.

Kont, A., J. Jaagus, R. Aunap, U. Ratas, and R. Rivis. 2008. "Implications of Sea-Level Rise for Estonia." *Journal of Coastal Research* 24 (2): 423–31.

Kysel, J., and R. Huth. 2004. "Heat-Related Mortality in the Czech Republic Examined through Synoptic and 'Traditional' Approaches." *Climate Research* 25: 265–74.

Lampietti, Julian A., David G. Lugg, Philip van der Celen, and Amelia Branczik. 2009. *The Changing Face of Rural Space: Agriculture and Rural Development in the Western Balkans.* Washington, DC: World Bank.

Lempert, R. J., and M. E. Schlesinger. 2000. "Robust Strategies for Abating Climate Change." *Climatic Change* 45 (3–4): 387–401.

Lewis, M. 2007. "In Nature's Casino." *New York Times Magazine*, August 26.

Ligeti, E., J. Penney, and I. Wieditz. 2007. *Cities Preparing for Climate Change: A Study of Six Urban Regions.* Toronto: Clean Air Partnership.

Luketina, D., and M. Bender. 2002. "Incorporating Long-Term Trends in Water Availability in Water Supply Planning." *Water Science and Technology* 46 (6–7): 113–20.

Mansoor, Ali, and Bryce Quillin, ed. 2007. *Migration and Remittances: Eastern Europe and the Former Soviet Union.* Washington, DC: World Bank.

Maracchi, G., O. Sirotenko, and M. Bindi. 2005. "Impacts of Present and Future Climate Variability on Agriculture and Forestry in the Temperate Regions: Europe." *Climatic Change* 70: 117–35.

Martinez, M. L., A. Intralawana, G. Vázquez, O. Pérez-Maqueo, P. Sutton, and R. Landgrave. 2007. "The Coasts of Our World: Ecological, Economic, and Social Importance." *Ecological Economics* 63 (2–3): 254–72.

Martinez-Baylach, J., A. Cabot Dalmau, L. Garcia Rodriguez, and G. Sauca. 2007. "Malaria importada: Revisión epidemiológica y clínica de una enfermedad emergente cada vez más frecuente." *Anales de pediatriia* 67 (3): 199–205.

McCarthy, J. J, O. F. Canziani, N. A. Leary, D. J. Dokken, and K. S. White, ed. 2001. *Climate Change 2001: Impacts, Adaptation, and Vulnerability. Contribution of Working Group II to the Third Assessment Report of the Intergovernmental Panel on Climate Change.* Cambridge, U.K.: Cambridge University Press.

Milly, P. C. D., J. Betancourt, M. Falkenmark, R. M. Hirsch, Z. W. Kundzewicz, D. P. Lettenmaier, and R. J. Stouffer. 2008. "Stationarity Is Dead: Whither Water Management?" *Science* 319: 573–74.

Milly, P. C. D., K. A. Dunne, and A. V. Vecchia. 2005. "Global Pattern of Trends in Streamflow and Water Availability in a Changing Climate." *Nature* 438: 347–50.

MRG (Minority Rights Group International). 2008. *State of the World's Minorities 2008.* London: MRG.

Myers, N., R. A. Mittermeier, C. G. Mittermeier, G. A. B. da Fonseca, and J. Kent. 2000. "Biodiversity Hotspots for Conservation Priorities." *Nature* 403: 853–58.

Næss, L. O., I. Thorsen Norland, W. M. Lafferty, and C. Aall. 2006 "Data and Processes Linking Vulnerability Assessment to Adaptation Decision-Making on Climate Change in Norway." *Global Environmental Change* 16: 221–33.

Nakat, Z. "Climate Change Adaptation in the Transport Sector." Background paper prepared for report, World Bank, Washington, DC.

National Geographic Society. 2001. Ecoregion Profile of Northeast Siberian Coastal Tundra. http://www.nationalgeographic.com/wildworld/profiles/terrestrial/pa/pa1107.html.

Natural Resources Canada. 2005. "An Overview of the Risk Management Approach to Adaptation to Climate Change in Canada." Paper prepared by

Global Change Strategies International, for the Climate Change Impacts and Adaptation Directorate of Natural Resources Canada, Ottawa.

New Zealand Climate Change Office. 2004. *Coastal Hazards and Climate Change: A Guidance Manual for Local Government in New Zealand.* Wellington: Ministry for the Environment.

Niederer, P., V. Bilenko, N. Ershova, H. Hurni, S. Yerokhin, and D. Maselli. 2008. "Tracing Glacier Wastage in the Northern Tien Shan (Kyrgyzstan/Central Asia) over the Last 40 Years." *Climatic Change* 86 (1–2): 227–34.

OECD (Organisation for Economic Co-operation and Development). 2005. *Financing Water Supply and Sanitation in Eastern Europe, Caucasus, and Central Asia.* Proceedings from a Conference of Eastern Europe, Caucasus, and Central Asia Ministers of Economy, Finance, and Environment and Their Partners, Yerevan, November 17–18.

OFDA (Office of United States Foreign Disaster Assistance). 1992. *Annual Report for Fiscal Year 1991.* Washington, DC: United States Agency for International Development.

———. 1998. *Annual Report for Fiscal Year 1997.* Washington, DC: United States Agency for International Development.

———. 1999. *Annual Report for Fiscal Year 1998.* Washington, DC: United States Agency for International Development.

Ogonesyan, V. 2004. "Regional Study of Hydromet Services in ECA Region: Damaging Transboundary Weather Systems." Consultant report prepared for the World Bank, World Bank, Washington, DC.

Olesen, J., and M. Bindi. 2002. "Consequences of Climate Change for European Agricultural Productivity, Land Use and Policy." *European Journal of Agronomy* 16: 239–62.

Oreskes, N. 2004. "Beyond the Ivory Tower: The Scientific Consensus on Climate Change." *Science* 306 (5702): 1686.

Padgham, J. 2008. "Report on Climate Change Adaptation in Agriculture." World Bank, Washington, DC.

Parry, M. L., C. Rosenzweig, and M. Livermore. 2005. "Climate Change, Global Food Supply and Risk of Hunger." *Philosophical Transactions of the Royal Society B: Biological Sciences* 360: 2125–38.

Parry, M., O. F. Canziani, J. P. Palutikof, P. J. van der Linden, and C. E. Hanson, ed. 2007. *Climate Change 2007: Impacts, Adaptation and Vulnerability. Contribution of Working Group II to the Fourth Assessment Report of the Intergovernmental Panel on Climate Change.* Cambridge, U.K.: Cambridge University Press.

Patrinos, A., and A. Bamzai. 2005. "Policy Needs Robust Climate Science." *Nature* 438: 285.

Pew Global Attitudes Project. 2007. *Rising Environmental Concern in 47-Nation Survey: Global Unease with Major World Powers.* Washington, DC: Pew Research Center.

Pollner, J., J. Kryspin-Watson, and S. Nieuwejaar. 2008. "Climate Change Adaptation in Europe and Central Asia: Disaster Risk Management." Background paper prepared for report, World Bank, Washington, DC.

Prasad, N., F. Ranghieri, F. Shah, Z. Trohanis, E. Kessler, and R. Sinha. 2009. *Climate Resilient Cities: A Primer on Reducing Vulnerabilities to Disasters.* Washington, DC: World Bank.

Price, M. F., and G. R. Neville. 2003. "Designing Strategies to Increase the Resilience of Alpine/Montane Systems to Climate Change." In *Buying Time: A User's Manual for Building Resistance and Resilience to Climate Change in Natural Systems,* ed. L. J. Hansen, J. L. Biringer, and J. R. Hoffman, 73–94. Berlin: World Wildlife Fund.

Pruszak, Z., and E. Zawadzka. 2008. "Potential Implications of Sea-Level Rise for Poland." *Journal of Coastal Research* 24 (2): 410–22.

Pusch, C. 2004. "Preventable Losses: Saving Lives and Property through Hazard Risk Management—A Comprehensive Risk Management Framework for Europe and Central Asia." Disaster Risk Management Working Paper 9, World Bank, Washington, DC.

Rabie, T., S. el Tahir, T. Alireza, G. Sanchez Martinez, K. Ferl, and N. Cenacchi. 2008. "The Health Dimension of Climate Change." Background paper prepared for report, World Bank, Washington, DC.

Randolph, T. F., E. Schelling, D. Grace, C. F. Nicholson, J. L. Leroy, D. C. Cole, M. W. Demment, A. Omore, J. Zinsstag, and M. Ruel. 2007. "Role of Livestock in Human Nutrition and Health for Poverty Reduction in Developing Countries." *Journal of Animal Science* 85: 2788–800.

Reid, H. 2006. "Climate Change and Biodiversity in Europe." *Conservation and Society* 4 (1): 84–101.

Renssen, H., B. C. Lougheed, J. C. J. H. Aerts, H. de Moel, P. J. Ward, and J. C. J. Kwadijk. 2007. "Simulating Long-Term Caspian Sea Level Changes: The Impact of Holocene and Future Climate Conditions." *Earth and Planetary Science Letters* 261: 685–93.

Repetto, R. 2008. "The Climate Crisis and the Adaptation Myth." Yale School of Forestry and Environmental Studies Working Paper 11, Yale University, New Haven, CT.

Revenga, A., D. Ringold, and W. M. Tracy. 2002. *Poverty and Ethnicity: A Cross-Country Study of Roma Poverty in Central Europe.* Washington, DC: World Bank.

Robine, J.-M., S. L. K. Cheung, S. Le Roy, H. Van Oyen, C. Griffiths, J.-P. Michel, and F. R. Herrmann. 2008. "Death Toll Exceeded 70,000 in Europe during Summer of 2003." *Comptes Rendus Biologies* 331 (2): 171–78.

Savoskul, O. S., E. V. Chevnina, F. I. Perziger, L. Y. Vasilina, V. L. Baburin, A. I. Danshin, B. Matyakubov, and R. R. Murakaev. 2003. "Water, Climate, Food, and Environment in the Syr Darya Basin." ADAPT Project Report, Tashkent.

Schiermeier, Q. 2004. "Modellers Deplore 'Short-Termism' on Climate." *Nature* 428: 593.

Schneider, S. H., and K. Kuntz-Duriseti. 2002. "Uncertainty and Climate Change Policy." In *Climate Change Policy: A Survey,* ed. S. H. Schneider, A. Rosencranz, and J. O. Niles, 53–88. Washington, DC: Island Press.

Scholze, M., W. Knorr, N. W. Arnell, and I. C. Prentice. 2006. "A Climate-Change Risk Analysis for World Ecosystems." *Proceedings of the National Academy of Sciences* 103 (35): 13116–20.

Schuur, E. A. G., J. Bockheim, J. G. Canadell, E. Euskirchen, C. B. Field, S. V. Goryachkin, S. Hagemann, P. Kuhry, P. M. Lafleur, H. Lee, G. Mazhitova, F. E. Nelson, A. Rinke, V. E. Romanovsky, N. Shiklomanov, C. Tarnocai, S. Venevsky, J. G. Vogel, and S. A. Zimov. 2008. "Vulnerability of Permafrost Carbon to Climate Change: Implications for the Global Carbon Cycle." *BioScience* 58 (8): 701–14.

Shkolnik, I. M., V. P. Meleshko, and V. M. Kattsov. 2007. "The MGO Climate Model for Siberia." *Russian Meteorology and Hydrology* 32 (6): 351–59.

Silander, J., B. Vehviläinen, J. Niemi, A. Arosilta, T. Dubrovin, J. Jormola, V. Keskisarja, A. Keto, A. Lepistö, R. Mäkinen, M. Ollila, H. Pajula, H. Pitkänen, I. Sammalkorpi, M. Suomalainen, and N. Veijalainen. 2006. "Climate Change Adaptation for Hydrology and Water Resources." FINADAPT Working Paper 6, Finnish Environment Institute Mimeographs 336, Helsinki.

Sinn, H.-W., G. Flaig, M. Werding, S. Munz, N. Düll, and H. Hofmann. 2001. *EU Enlargement and Labor Migration: Towards a Gradual Convergence of the Labor Market.* Munich: CESifo Group.

Sirohi, S., and A. Michaelowa. 2007. "Sufferer and Cause: Indian Livestock and Climate Change." *Climatic Change* 85 (3–4): 285–98.

Sirotenko, O. D., H. V. Abashina, and V. N. Pavlova. 1997. "Sensitivity of the Russian Agriculture to Changes in Climate, CO_2, and Tropospheric Ozone Concentrations and Soil Fertility." *Climatic Change* 36 (1–2): 217–32.

Smith, J. B., S. H. Schneider, M. Oppenheimer, G. W. Yohe, W. Hare, M. D. Mastrandrea, A. Patwardhan, I. Burton, J. Corfee-Morlot, C. H. D. Magadza, H.-M. Füssel, A. Barrie Pittock, A. Rahman, A. Suarez, and J.-P. van Ypersele. 2009. "Assessing Dangerous Climate Change through an Update of the Intergovernmental Panel on Climate Change (IPCC) 'Reasons for Concern.'" *Proceedings of the National Academy of Sciences* 106 (11): 4133–37. http://www.pnas.org/content/early/2009/02/25/0812355106 .full.pdf+html

Sokolov, A. P., P. H. Stone, C. E. Forest, R. Prinn, M. C. Sarofim, M. Webster, S. Paltsev, C. A. Schlosser, D. Kicklighter, S. Dutkiewicz, J. Reilly, C. Wang, B. Felzer, and H. D. Jacoby. 2009. "Probabilistic Forecast for 21st Century Climate Based on Uncertainties in Emissions (without Policy) and Climate Parameters." Report 169, Massachusetts Institute of Technology Joint Program on the Science and Policy of Global Change, Cambridge, MA, January.

Solomon, S., D. Qin, M. Manning, Z. Chen, M. Marquis, K. B. Averyt, M. Tignor, and H. L. Miller, ed. 2007. *Climate Change 2007: The Physical Science Basis. Contribution of Working Group I to the Fourth Assessment Report of the Intergovernmental Panel on Climate Change.* Cambridge, U.K.: Cambridge University Press.

Solomon, S., G.-K. Plattner, R. Knutti, and P. Friedlingstein. 2009. "Irreversible Climate Change Due to Carbon Dioxide Emissions." *Proceedings of the National Academy of Sciences* 106 (6): 1704–9. http://www.pnas.org/content/ early/2009/01/28/0812721106.short.

Somlyódy, L. 2002. *Strategic Issues of the Hungarian Water Resources Management.* Budapest: Academy of Science of Hungary.

Sutton, W. R., R. I. Block, and J. Srivastava. 2008. "Adaptation to Climate Change in Europe and Central Asia Agriculture." Background paper prepared for report, World Bank, Washington, DC.

Sutton, W. R., P. Whitford, E. Montanari Stephens, S. Pedroso Galinato, B. Nevel, B. Plonka, and E. Karamete. 2007. *Integrating Environment into Agriculture and Forestry: Progress and Prospects in Eastern Europe and Central Asia,* vol 1. Washington, DC: Europe and Central Asia Region, Sustainable Development Department, World Bank.

Swinnen, J., and S. Rozelle. 2006. *From Marx and Mao to the Market: The Economics and Politics of Agrarian Transition.* Oxford, U.K.: Oxford University Press.

Taber, K. C. 2008. "Fight Climate Change by Turning Green." International Health Tribune, March 19.

Taleb, N. N. 2007. *The Black Swan: The Impact of the Highly Improbable.* New York: Random House.

Turner, B. L., R. E. Kasperson, P. A. Matsone, J. J. McCarthy, R. W. Corell, L. Christensen, N. Eckley, J. X. Kasperson, A. Luers, M. L. Martello, C. Polsky, A. Pulsipher, and A. Schiller. 2003. "A Framework for Vulnerability Analysis in Sustainability Science." *Proceedings of the National Academy of Sciences* 100 (14): 8074–79.

UKCIP (United Kingdom Climate Impacts Programme). 2003. *Climate Adaptation: Risk, Uncertainty and Decision-making.* Technical Report, UKCIP, Oxford, U.K.

UNDP (United Nations Development Program). UNDP Country Disaster Risk Index (DRI) Tool. http://gridca.grid.unep.ch/undp/.

UNDP-Albania and Ministry of Environment. 2002. "The First National Communication of the Republic of Albania to the UNFCCC." UNDP-Albania, Tirana.

UNDP-Turkey and Ministry of Environment and Forestry. 2007. "First National Communication on Climate Change." UNDP-Turkey, Ankara.

UNHCR (United Nations High Commissioner for Refugees). 2004. *Profile of Internal Displacement: Georgia. Compilation of the Information Available in the Global IDP Database of the Norwegian Refugee Council.* Geneva: UNHCR.

UNISDR (United Nations International Strategy for Disaster Risk Reduction) and World Bank. 2007. "South Eastern Europe Disaster Risk Management Initiative: Desk Study Review Risk Assessment in South Eastern Europe." United Nations, Geneva.

UN Moldova 2007a. "Relief and Technical Assistance Response to the Drought in Moldova." United Nations Moldova, Chisnau.

UN Moldova 2007b. "Relief and Technical Assistance Response to the Drought in Moldova—Project Status Report No. 1." United Nations Moldova and Government of Moldova, Chisnau.

UN OCHA (United Nations Office for the Coordination of Humanitarian Affairs). 2008. "Floods in Kazakhstan." Situation Report 1, March. Almaty, Kazakhstan, UN OCHA.

Valiela, I. 2006. *Global Coastal Change.* Oxford, U.K.: Blackwell.

van der Heijden, K. 1996. *Scenarios: The Art of Strategic Conversation.* West Sussex, U.K.: John Wiley & Sons.

Vörösmarty, C. J., P. Green, J. Salisbury, and R. B. Lammers. 2000. "Global Water Resources: Vulnerability from Climate Change and Population Growth." *Science* 289: 284–88.

Wang, Y., L. R. Leung, J. L. McGregor, D. K. Lee, W. C. Wang, Y. Ding, and F. Kimura. 2004. "Regional Climate Modeling: Progress, Challenges, and Prospects." *Journal of the Meteorological Society of Japan* 82: 1599–628.

Weitzman, M. 2009. "On Modeling and Interpreting the Economics of Catastrophic Climate Change." *Review of Economics and Statistics* 91 (1): 1–19.

Westphal, M. I. 2008. "Summary of the Climate Science in the Europe and Central Asia Region: Historical Trends and Future Projections." Background paper prepared for report, World Bank, Washington, DC.

WHO (World Health Organization) Europe. 2001. "Health Aspects of the Drought in Uzbekistan 2000–2001." Technical Field Report Series, July. WHO Regional Office for Europe, Emergency Preparedness and Response Program, Copenhagen, Denmark.

———. 2002. "Epidemiological Surveillance of Malaria in Countries of Central and Eastern Europe and Selected Newly Independent Status." Report on a WHO intercountry meeting, Sofia, June.

———. 2003. *Guidelines for Safe Recreational Water Environments, Volume 1: Coastal and Fresh Waters.* Geneva: World Health Organization.

———. 2005. "Inception Meeting on the Malaria Elimination Initiative in the WHO European Region: Report on a WHO meeting, Tashkent, Uzbekistan, October 18–20, 2005." WHO Regional Office for Europe, Copenhagen, Denmark.

WHO (World Health Organization) and UNICEF (United Nations Children's Fund). 2006. *Meeting the MDG Drinking Water and Sanitation Target: The Urban and Rural Challenge of the Decade.* Geneva: WHO and UNICEF.

Wilby, R. L., J. Troni, Y. Biot, L. Tedd, B. C. Hewitson, D. G. Smith, and R. T. Sutton. 2009. "A Review of Climate Risk Information for Adaptation and Development Planning." *International Journal of Climatology* 29 (9): 1193–215.

WMO (World Meteorological Organization). 2007. "2007 Drought in the Republic of Moldova." http://www.wmo.int/pages/publications/meteoworld07/_archive/en/october2007/moldavia_drought.html.

World Bank. 2001. "Mozambique Flood Emergency Recovery Fund." Technical Annex P070432. World Bank, Washington, DC.

———. 2003. *Water Resources in Europe and Central Asia—Volume 1: Issues and Strategic Directions.* Washington, DC: World Bank.

———. 2005. *Drought Management and Mitigation Assessment for Central Asia and the Caucasus.* Washington, DC: World Bank.

———. 2006. *Drought Management and Mitigation Assessment for Central Asia and the Caucasus: Regional and Country Profiles and Strategies.* Washington, DC: World Bank.

———. 2007. *World Development Report 2008: Agriculture for Development.* Washington, DC: World Bank.

———. 2008. World Development Indicators. Washington, DC: World Bank.

WWF (World Wide Fund For Nature). 2008. "Lena River Delta." http://www.panda.org/about_wwf/where_we_work/ecoregions/lena_river_delta.cfm.

Yohe, G. W., R. D. Lasco, Q. K. Ahmad, N. W. Arnell, S. J. Cohen, C. Hope, A. C. Janetos, and R. T. Perez. 2007. "Perspectives on Climate Change and Sustainability." *Climate Change 2007: Impacts, Adaptation, and Vulnerability—Contribution of Working Group II to the Fourth Assessment Report of the Intergovernmental Panel on Climate Change*, ed. M. L. Parry, O. F. Canziani, J. P. Palutikof, P. J. van der Linden, and C. E. Hanson, 811–42. Cambridge, U.K.: Cambridge University Press.

Zhang, X., and J. E. Walsh. 2006. "Toward a Seasonally Ice-Covered Arctic Ocean: Scenarios from the IPCC AR4 Model Simulations." *Journal of Climate* 19: 1730–47.

Index